In-Cell NMR Spectroscopy

In-Cell NMR Spectroscopy: Biomolecular Structure and Function

Special Issue Editors

Alexander Shekhtman

David S. Burz

MDPI • Basel • Beijing • Wuhan • Barcelona • Belgrade

Special Issue Editors

Alexander Shekhtman
State University of New York
USA

David S. Burz
State University of New York
USA

Editorial Office
MDPI
St. Alban-Anlage 66
4052 Basel, Switzerland

This is a reprint of articles from the Special Issue published online in the open access journal *International Journal of Molecular Sciences* (ISSN 1422-0067) from 2018 to 2019 (available at: https://www.mdpi.com/journal/ijms/special_issues/in_cell_nmr).

For citation purposes, cite each article independently as indicated on the article page online and as indicated below:

LastName, A.A.; LastName, B.B.; LastName, C.C. Article Title. *Journal Name* **Year**, *Article Number*, Page Range.

ISBN 978-3-03928-254-8 (Pbk)
ISBN 978-3-03928-255-5 (PDF)

Cover image courtesy of Alexander Shekhtman and David S. Burz.

Contents

About the Special Issue Editors

Alexander Shekhtman completed his Ph.D. in NMR spectroscopy at the State University of New York at Albany and his postdoctoral studies at Rockefeller University and New York Structural Biology Center in New York. He is a Full Professor in the Department of Chemistry at the State University of New York at Albany, has published more than 100 papers in peer-reviewed journals, and is a series editor of Methods in *Molecular Biology* and a member of the Scientific Reports Editorial Board. His interests lie in developing in-cell NMR technology and applying this technology to study molecular interactions within biologically relevant systems.

David S. Burz completed his Ph.D. in Molecular Biology and Biochemistry from Wesleyan University in Connecticut, and post-doctoral studies at Washington University School of Medicine in St. Louis, Wadsworth Center New York State Department of Health in Albany, and the State University of New York at Albany. He was an Assistant Professor at the Albany College of Pharmacy and is currently an Instructional Specialist in the Department of Chemistry at the State University of New York at Albany. He has published more than 50 peer-reviewed papers. His interests center on how linked functions resulting from macromolecular interactions contribute to the regulation of biological activity.

Preface to "In-Cell NMR Spectroscopy: Biomolecular Structure and Function"

One of the fundamental questions of scientific inquiry is to understand how the emergent property of life results from a near limitless array of molecular interactions. Traditional benchtop methodologies used to investigate these interactions do not provide the appropriate milieu for answering this fundamental question due to the extreme complexity of physiological states. The interior of a cell is an especially dense environment containing up to 400 mg/mL of bio-molecular species and reduced amounts of bulk water. Hydrophilic, hydrophobic and electrostatic interactions behave differently than in dilute solutions consisting of limited amounts of purified components. The effects of excluded volume and molecular crowding increase the concentration of cytosolic species, assuring the likelihood of transient low-affinity interactions. The ability to elucidate molecular structures and interaction dynamics under such conditions has long been the goal of cellular and molecular structural biologists. In-cell NMR spectroscopy brings us closer to realizing this goal by providing atomic level resolution of the molecules engaged in physiologically relevant interactions within the complex interior of a living cell.

Over the past 20 years, the field of in-cell NMR spectroscopy has evolved from proof-of-principle to methodologies with a wide spectrum of applications. This Special Issue presents a conspectus of research and recent innovations from prominent laboratories in the field of solid state and solution in-cell NMR spectroscopy. Chapters 1 and 2 summarize the salient aspects of performing in-cell NMR spectroscopy. Chapter 1 reviews the types of studies that have been performed to date, while chapter 2 focuses on methodologies developed to study protein-small molecule interactions and post-translational modifications. Special emphasis in chapter 2 is on studies of metabolism by using hyperpolarization NMR methods, which enhance the sensitivity of NMR spectroscopy by several orders of magnitude. Chapter 3 is a perspective of bio-molecular NMR spectroscopy and an outline of methodologies that expand future in-cell NMR studies with an emphasis on time-resolved solution NMR spectroscopy to examine post-translational modifications and an initiative to assess the effect of cellular constituents on protein quinary structures. Chapter 4 reviews the current state of structural analyses of proteins inside living cells, discussing the potential of structure determination to advance structure-based drug screening and to augment the role of structure in assessing biological activity and its regulation. Chapter 5 examines the structure of the intrinsically disordered Tau protein and its interactions with cytoskeletal components in mammalian cells as well as post-translational modifications. Intriguingly, a sophisticated post-translational modification system may operate to maintain proper post-translational modification and correct abnormal modifications. Chapter 6 summarizes the effect of ribosomes on the quality of in-cell NMR protein spectra in prokaryotes and ascribes these effects to quinary interactions that also alter the activity of the target molecules. Chapters 7 and 8 address advances in solid state NMR spectroscopy of biological molecules. Chapter 7 uses 31P NMR spectroscopy of nuclei acids and phospholipids to evaluate the integrity of bacterial cells in response to antimicrobial agents Chapter 8 investigates the structure of starch in microalgal cells and its accumulation/degradation in response to metabolic stress.

The body of work presented illuminates the future of in-cell NMR spectroscopy. Experiments are now capable of determining the structure of biological macromolecules within a living cell, discerning time resolved structural modifications and interactions that regulate biological processes, and monitoring the metabolic response of cells to various stimuli. The combination of information gathered from these studies will advance the perception of how the intricate network of interactions within cells gives rise to the emergent property of life.

Alexander Shekhtman, David S. Burz

Special Issue Editors

International Journal of
Molecular Sciences

Review

Applications of In-Cell NMR in Structural Biology and Drug Discovery

CongBao Kang

Experimental Therapeutics Centre, Agency for Science, Technology and Research (A*STAR), 31 Biopolis way, Nanos, #03-01, Singapore 138669, Singapore; cbkang@etc.a-star.edu.sg; Tel.: +65-6407-0602

Received: 26 November 2018; Accepted: 29 December 2018; Published: 2 January 2019

Abstract: In-cell nuclear magnetic resonance (NMR) is a method to provide the structural information of a target at an atomic level under physiological conditions and a full view of the conformational changes of a protein caused by ligand binding, post-translational modifications or protein–protein interactions in living cells. Previous in-cell NMR studies have focused on proteins that were overexpressed in bacterial cells and isotopically labeled proteins injected into oocytes of *Xenopus laevis* or delivered into human cells. Applications of in-cell NMR in probing protein modifications, conformational changes and ligand bindings have been carried out in mammalian cells by monitoring isotopically labeled proteins overexpressed in living cells. The available protocols and successful examples encourage wide applications of this technique in different fields such as drug discovery. Despite the challenges in this method, progress has been made in recent years. In this review, applications of in-cell NMR are summarized. The successful applications of this method in mammalian and bacterial cells make it feasible to play important roles in drug discovery, especially in the step of target engagement.

Keywords: in-cell NMR; protein structure; protein dynamics; drug discovery; target engagement; protein modification

---

1. Introduction

Solution nuclear magnetic resonance (NMR) [1], X-ray crystallography and cryogenic electron microscopy (cryo-EM) [2] are important tools for obtaining the structures of biomolecules at atomic resolution [3]. When diffracted crystals are available, X-ray crystallography is a robust way to obtain high-resolution structures of biomolecules [4]. In recent years, the rapid development of cryo-EM has made it possible to solve structures of biomolecule complexes with high molecular weight at a high resolution. For example, the structures of many difficult targets such as ion channels and membrane-bound enzyme complexes were obtained using cryo-EM [5,6]. Other methods, such as small-angle X-ray/neutron scattering (SAXS/SANS) [7], mass spectrometry [8] and chemical cross-linking [9] are also used to determine structures of protein complexes.

Solution NMR spectroscopy is able to investigate protein structures and dynamics under solution conditions because the targets can be studied in different buffers and at various temperatures [10]. Although it is still challenging to study protein structures with high molecular mass due to the signal overlap and sensitivity, NMR has been widely used in protein chemistry and drug discovery with the development of magnets, pulse programs [11–13], and different protein-labeling strategies [14–16]. Solution NMR spectroscopy has been used in various research topics, including protein–protein, protein–nucleotide complexes, and membrane proteins, to provide useful information in order to understand protein structure and function [17–20]. Both solid and solution NMR spectroscopies have been successfully used to probe the structures of membrane proteins, which are normally challenging to crystallize [21–23]. Many membrane proteins have been characterized using solution and solid-state NMR spectroscopy [24–26].

NMR spectroscopy is a powerful method that can be used in combination with other methods, such as X-ray, cryo-EM, bioinformatics and SAXS/SANS, providing different views on the structures and dynamics of biomolecules, and their functional complexes in solution [27–31]. It is well known that NMR data analysis is time consuming. Therefore, NMR can work with other methods to save a lot of time in data processing and analysis. Available web servers, such as structure prediction and protein–protein binding interface predictions, can also speed up NMR data analysis [32–39]. The most frequently used strategy is to combine available structures obtained by using X-ray, cryo-EM or homology models with dynamic and ligand binding information obtained by NMR, which provides a full view of the target function, ligand binding modes, and regulation mechanisms [27]. Successful examples can be seen in many studies [40–43], and will not be described here.

As NMR is a powerful tool for monitoring the environmental changes of atoms, it has been used for probing protein–protein and –ligand interactions. In addition, NMR active nuclei such as ^{19}F and ^{31}P can be incorporated to a protein, making ^{19}F and ^{31}P NMR possible in determining conformational changes of proteins induced by ligand binding or post-translational modifications [44–48]. In fragment-based drug discovery (FBDD), NMR is frequently used in identifying fragments with different binding affinities [49,50]. Proton-based NMR spectroscopies have been successfully used in this field. As hetero-nuclear NMR experiments can be used to monitor environmental changes of individual amino acid of a protein, NMR is then very useful in generating the structure-activity relationship of a compound in a drug discovery project [47,51]. The available access to different types of compound libraries such as ^{19}F-labeled compound libraries makes NMR an important tool in drug discovery by identifying novel hits, confirming hits obtained from biochemical assays, mapping the ligand binding site, probing the druggability of a target protein, and determining the ligand binding mode [45,46,48,52–55].

With the accumulation of structures of biomolecules determined by different methods such as X-ray and Cryo-EM, interest has been focused on the correlation between structure and function of biomolecules. Therefore, the information obtained from structural biology has to be connected well with that obtained from cell biology and biochemistry. It is critical that the structure of a biomolecule is determined under a condition that is close to the physiological environment. NMR is the most efficient structural tool to achieve such requirements [56]. Research has been carried out to study the structures of proteins in living cells using NMR techniques, which leads to the concept of in-cell NMR [57]. This unique approach bridges the gap between structural techniques and cellular imaging techniques [58,59]. This technique is also applicable to solid-state NMR [60,61]. This review only summarizes recent progress in in-cell NMR using solution NMR spectroscopy and discusses the challenges and potential applications in drug discovery.

2. In-Cell NMR

In-cell NMR was proposed to study protein dynamics and structures in living cells [62], making this method unique to others used for structural analysis [57,63]. It is a non-invasive method to determine the structure of a target under the physiological conditions [64]. As the cells used in in-cell NMR are alive, intact and contain complete cellular compartments, the obtained information is therefore very useful in biology, as well as other fields, such as drug discovery. Although structural studies of membrane proteins in living cells are of great interest for in-cell NMR, this review will mainly focus on in-cell NMR studies of water-soluble proteins carried out using solution-state NMR [61,65].

2.1. Cells Used in In-Cell NMR

Different cells, including bacteria, yeast, oocyte and mammalian cells, are able to be used for in-cell NMR studies. The most frequently used cell line is *E. coli* (Tables 1 and 2). The application of in-cell NMR in mammalian cells make it attractive in target engagement in drug discovery when the targets are related to human diseases. It will be ideal when in-cell NMR can be carried out in all types of cells, while experiments have to be performed to obtain suitable conditions for gaining high-quality NMR spectra.

Table 1. Some types of experiment used in in-cell NMR studies [a].

Experiment	Remarks	Reference
^{1}H-^{15}N-HSQC (heteronuclear single quantum coherence)	Protein–protein/ligand interactions	[66,67]
3D experiments	Backbone assignment	[68]
PCS (pseudo-contact shift)	Protein structure determination using lanthanide tags	[69,70]
NOESY (Nuclear Overhauser effect spectroscopy)	Protein structure determination	[71]
SOFAST-HMQC (Band-Selective Optimized Flip Angle Short Transient- heteronuclear multiple quantum coherence)	Protein–protein/ligand interactions	[72]
^{1}H-^{13}C HSQC	Protein structure analysis using selectively protonation and ^{13}C labeling	[68]
^{19}F-NMR	In-cell protein-observed ^{19}F can be obtained	[73]
Relaxation	Protein dynamics	[74]
Residue dipolar couplings	Lanthanide tags can also be used to generate RDCs	[69]
Protein-based-^{1}H NMR	^{1}H-NMR at His residue regions	[75]
Ligand-based ^{1}H NMR	Protein-ligand interactions	[76]
^{19}F-NMR	Ligand observed ^{19}F-NMR was used in ligand binding studies	[77]

[a] Not all the references are listed in the table for the same type of experiments.

Table 2. In-cell NMR studies of proteins in different cells.

Cells	Targets	Studies	Reference
	TTHA1718	Structure was determined in the living cells	[68]
	calmodulin, NmerA, and FKBP (FK506 binding protein)	Labeling methyl groups of protein was used in-cell NMR studies	[78]
	HdeA, alpha-synuclein, chymotrypsin inhibitor 2 (CI2) ubiquitin	Protein dynamics in cells, protein leakage, and protein–protein interactions were analyzed	[63,79,80]
	Thioredoxin ADK (adenosine kinase) FKBP	Quandary interactions of proteins in cells was addressed in the study	[81]
Bacteria	Alpha-synuclein, ubiquitin, HDH (histidinol dehydrogensase), GFP (Green fluorescence protein)	Protein-based ^{19}F-NMR study was carried out	[73]
	SOD1 SOD1 (human copper, zinc superoxide dismutase 1)	Protein folding in living cells was analyzed.	[72]
	PFN1 (protein profilin 1)	Protein–protein interaction was studied in living cells	[82]
	Pup (prokaryotic ubiquitin like protein) Mpa (mycobacterial protease ATPase)	In-cell NMR was used to screen compounds disrupting protein–protein interactions	[83]
	FKBP12	In-cell NMR was used to screen a library.	[84]
	Cox17 (cytochrome c oxidase copper chaperone)	In-cell NMR was used to probe protein folding in living cells	[85]

Table 2. *Cont.*

Cells	Targets	Studies	Reference
oocyte	Ubiquitin, calmodulin	Protein–protein interactions were probed in oocyte	[86]
	GB1 (the B domain of G protein)	Structural studies were performed using PRE restrains	[70,87,88]
	XT-GB1 (SV40 regulatory domain-GB1)	Protein phosphorylation was monitored in cells	[89]
yeast	Ubiquitin	Structural studies were carried out in cell compartments	[90]
Insect	GB1, HB8 TTHA1718, rat calmodulin, and human HAH1	3D experiments were collected in living insect cells for structural studies.	[71]
Mammalian cells	Tβ4 (thymosin β4)	Introducing proteins into cells using toxin was used for in-cell NMR studies.	[91]
	Thioredoxin	Redox status of intracellular thioredoxin was measured in living cells	[92]
	GB1 FKBP12	Labeled protein was delivered into mammalian cells using peptides for in-cell NMR	[93]
	Alpha-synuclein	Protein modification and folding were monitored	[94,95]
	hSOD1 and mutants	Folding in living cells and protein–protein interactions were analyzed	[96,97]
	SOD1	Effect of ebselen and ebsulphur on protein structure was investigated	[98]
	Mia40 (mitochondrial intermembrane space import and assembly protein 40)	Protein folding in living cells was investigated	[99]
	Cox17	Protein folding was investigated in living cells	[100]
	DNA i-motif	Stability of DNA i-motif was investigated.	[101]
	copper binding protein HAH1	Sequential protein expression in mammalian cells and selective labeling proteins was used in-cell NMR studies	[102]
	DJ1	Protein folding was investigated	[103]
	Bcl-2 (B-cell lymphoma 2)	Protein-ligand interactions. Saturation-Transfer Difference (STD) and TrNOE experiments were carried out	[104]
	PFN1	Specific and unspecific interactions in cells was explored using in-cell NMR	[82]

2.2. Isotopic Incorporation

Similar to the conventional NMR methods, to obtain high-quality in-cell NMR spectra, the proteins need to be isotopically labeled or contain NMR-active nuclei such as ^{15}N and ^{13}C. Labeling protein with ^{19}F [105] or ^{31}P is also a feasible strategy for in-cell NMR experiments, as ^{19}F and ^{31}P [106] NMR are commonly used in solution NMR studies. In-cell NMR has another advantage over other methods used for structural studies. Purifying the target protein is not required, which is very attractive for some targets that are difficult to prepare in vitro. Isotopically labeled proteins can be purified for in-cell NMR studies in mammalian cells, but they must be delivered to the cells (Figure 1) using cell-penetrating peptides, toxin microinjection, or electroporation methods [87,91,93]. Overexpressing the target proteins by growing cells in different medium is the most convenient way for in-cell NMR studies, which is achievable in both bacterial and mammalian cells (Table 2).

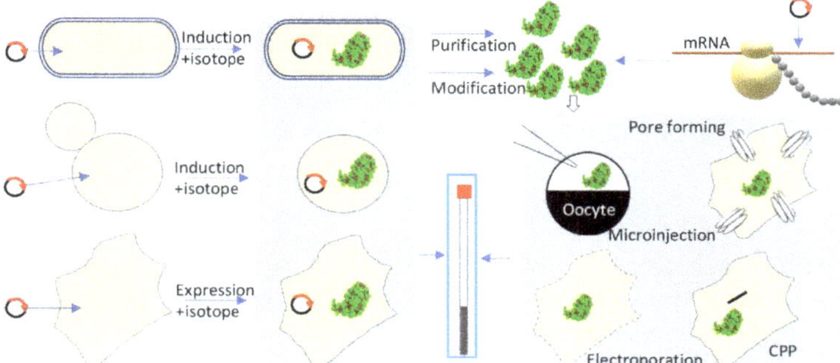

Figure 1. Sample preparation for in-cell NMR studies. The cells used for in-cell NMR studies can be prepared using the following strategies: Proteins (green) can be directly over-expressed in different cell lines using expression vectors. To make isotopically labeled proteins for in-cell NMR studies, the target gene can be cloned into suitable vectors followed with transfection/transformation into cells. Target protein can be isotopically labeled by growing cells in isotopically enriched (^{15}N, ^{15}N/^{13}C) media. Cells with the overexpressed protein are then used for in-cell NMR experiments. Isotopically labeled proteins can also be prepared in vitro by overexpressing them in different cells or using cell-free expression systems. The labeled protein is then purified before being introduced to oocytes by microinjection. Blue box indicates the NMR tube. Labeled proteins can also be introduced into human cells using either cell-penetrating peptides (CPP), cell permeabilization by pore-forming toxins or electroporation as introduced previously [107]. This figure was modified from the figure of Luchinat and Banci [107].

2.3. NMR Experiments for In-Cell NMR Studies

Although in-cell NMR experiments are similar to normal experiments that are carried out in vitro, several factors (below) will affect the selection of the experiments as challenges remain in the in-cell NMR studies. Normal one dimensional (1D) and multiple dimensional experiments can be collected (Table 1). As the available in-cell NMR studies focused on a few proteins (Table 2), more studies are needed to enlarge the application of this method.

2.4. Challenges in In-Cell NMR

Challenges remain for in-cell NMR in practice despite recent progress. Firstly, the target is present with other molecules in cells, which requires careful protein-labeling strategies to reduce the background signals. Secondly, the target protein might interact with other proteins to form complexes with high molecular weights which have rapid relaxation and low signal sensitivity. Unspecific interactions may also exist inside living cells, which will contribute to signal reduction in the NMR spectra. Optimized NMR pulse programs will be helpful in increasing the signal sensitivity and reducing data collection time [108,109]. Thirdly, injecting or delivering isotopically enriched proteins into cells is a good strategy for gaining signal intensity and reducing background noise, but the injected protein might be transported outside of the cells by different mechanisms. The leaky protein will exhibit signals influencing the in-cell NMR spectra [79]. When the protein is overexpressed in cells, the collapse of the dead cells will make the labeled protein released into the medium, which will give very sharp signals in the spectra [63,110]. A bioreactor in NMR tube can reduce cell death and make in-cell NMR sample last for a longer time [111,112]. Fourthly, the viscosity inside cells is higher than water, which can lead to line broadening of the signals [113,114]. Fifthly, the target protein may exist in different forms when it is over-expressed in the cells. The target protein might be in free form, in complexes with molecules and partially degraded by proteases. Such sample heterogeneity will give rise to in-cell NMR data with low quality. Sixthly, in ligand-binding studies, the tested ligands must be able to penetrate the cell membrane, which is different from the in vitro NMR study. The tested compounds should fulfill certain standards such as stability and cell penetrating activity when they are used in in-cell NMR experiments. Lastly, as in-cell NMR is monitoring spectra of a protein in living cells, the time required for data acquisition should be as short as possible because the target protein might be degraded by proteases. Strategies such as increasing protein stability, sustaining the life time of the cells, collecting data in a shorter time and using multiple samples in data collection will be helpful in in-cell NMR studies.

3. In-Cell NMR in Different Cells

In-cell NMR has been carried out in various cells (Table 2). Although most experiments are 1D and 2D types, accumulated studies (Table 3) provide evidence that other multiple dimensional experiments could be performed in various cell types.

Table 3. Some representative in-cell NMR studies.

System	Experimental Outcome	Reference
E. coli	Heteronuclear spectra of proteins were collected in living cells	[66]
E. coli	Protein structure was determined in living cells	[68]
Mammalian cells	In-cell NMR study of proteins that were delivered into cells was performed	[93]
Oocyte	Lanthanide tag was used in generating distance restraints in living cells	[115]
HEK293T	Protein was overexpressed in mammalian cells for in-cell NMR studies	[116]
E. coli	In-cell NMR was used to screening a library	[84]
M. smegmatis	The first application of in-cell NMR in target engagement	[117]
Hela	In-cell NMR study on DNA was carried out	[101]

3.1. In-Cell NMR in Bacterial Cells

In normal NMR samples, the concentration of the target protein is in the µM to mM range, with high purity (>90%). The concentration of a target protein in the living cells is normally very low, and there are a lot of proteins that might exhibit detectable NMR signals. The background signals from other molecules are very high if the cells are cultured in a medium containing isotopically labeled carbon and nitrogen sources. Overexpression of the target protein in the living cells is a strategy to gain signal intensities while the expression of other proteins should be properly suppressed.

To reduce the signal background of *E. coli* proteins, the following method can be used. The gene of a target protein cloned in an expression vector is first transformed into *E. coli* followed by culturing in the normal medium. Before the target protein was induced, the cultured bacterial cells were transferred into a medium containing isotopes [68], which reduced the background signals. This method was successfully used in the study of the putative heavy-metal binding protein TTHA1718. In the study, the sample was shown to be stable for 6 h. Backbone resonance assignment of the protein in cells were obtained using 3D experiments, which were collected using a nonlinear sampling scheme for the indirectly acquired dimensions [68]. In addition, selective protonation and ^{13}C labeling of Ala, Leu and Val residues of the protein were obtained in *E. coli*, which made structural determination of TTHA1718 in *E. coli* possible. This study showed the structure of the protein in the living cells. Although the structure in vivo is similar to that determined in vitro, residues that interact with other proteins can be identified. Isotopic labeling of the protein can also be achieved by switching cells from unlabeled medium to an isotope enriched medium [78]. This method can also be used for labeling protein at the methyl groups [78].

Most proteins might not be suitable for in-cell NMR studies [118], which makes in-cell NMR in *E. coli* cells only applicable to some specific cases. In addition to TTHA1718, several proteins, such as NumerA [66], GB1, the N-terminal metal-binding domain of MerA [119] and human copper, zinc superoxide dismutase 1 (hSOD1) [72], were shown to exhibit nicely dispersed cross peaks in the spectra in in-cell NMR studies (Table 2). For the folded proteins, the difficulty in obtaining good quality NMR data is mainly due to crowding [120]. For mammalian proteins, *E. coli* might not be an ideal system for in-cell NMR studies and the mammalian cells should be considered [120]. In-cell NMR study on some intrinsically disordered proteins can be carried out in *E. coli* cells using an overexpression system [121]. The procedures for carrying out such experiments have been described in detail [88,121]. In-cell NMR in bacteria is a powerful tool to evaluate structure and dynamics of intrinsically disordered proteins [63,122,123]. Protein-based ^{19}F-NMR was able to be carried out in *E. coli*, making it possible with this method to monitor proteins with high molecular weight [73]. Measuring the spin relaxation parameters was used to probe the interactions of intrinsically disordered protein and components of the cytosol in the living cells [74]. The dynamic parameters of intrinsically disordered proteins obtained using in-cell NMR under the physiological conditions will be useful for understanding their function and regulation [124].

3.2. In-Cell NMR in Yeast

Yeast cells such as *Pichia pastoris* are suitable for in-cell NMR studies, as they are used for overexpressing proteins in vitro NMR studies. For some mammalian proteins that are difficult to express in bacteria, yeast cells would be one option for protein production. In vitro NMR experiments demonstrated the interactions between ubiquitin and RNA in yeast [125]. Such interaction could be verified by in-cell NMR in yeast. A protocol for isotopic labeling of proteins in budding yeast was developed [90]. Ubiquitin was overexpressed using the *AOX1* promoter, which was induced by methanol. Ubiquitin in yeast cells was isotopically labeled and exhibited a dispersed NMR spectrum. The dynamic properties of ubiquitin in various cellular compartments, including cytosol and protein storage bodies, were explored using in-cell NMR. One advantage of using yeast in in-cell NMR studies is that the location of the overexpressed ubiquitin at different places were able to be achieved by growing cells in different growth media [90]. The impact of a target protein at different locations in living cells can therefore be investigated.

3.3. In-Cell NMR in Oocytes of Xenopus laevis

Oocyte was able to serve as a system for in-cell NMR studies in which microinjection of labeled proteins into the living cells was required [86]. As the size of the oocyte is larger than those of bacteria and mammalian cells, the amount of the cells in the NMR studies is less. Approximately 200 oocytes would be sufficient for one NMR measurement [87]. The cellular environment of the oocyte is close to that of the mammalian cells, which makes it a useful system to explore structure and function of human proteins [126,127]. To carry out in-cell NMR studies in oocytes, the target protein needs to be isotopically labeled, purified and then introduced into cells by microinjection. Several examples have proven the feasibility of this method. In a study carried out by Sakai et al., ^{1}H-^{15}N-HSQC spectrum of ubiquitin was obtained. Slightly different spectra of ubiquitin in cells and in vitro were observed. The amino acids that exhibited different chemical shifts in the spectra might be due to unspecific protein–protein interactions. In addition, maturation of ubiquitin precursor in the living cells was observed [86]. NMR studies of GB1 were also able to be carried out in oocytes [87]. In this study, purified GB1 was shown not to interact with any components of *Xenopus* egg extracts. The impact of BSA on the NMR spectra of GB1 was also investigated, which proves that oocytes can serve as a system for structural and binding studies on human proteins due to their possessing a similar environment to that found in human cells [87]. Using this approach, lanthanide-labeled proteins were able to be injected into oocyte. Distance restraints such as PCSs [115] and paramagnetic residual dipolar couplings (RDCs) [128] can be obtained, which can be utilized for determining protein structures and monitoring conformational changes. This method has been successfully used for structural studies on GB1 protein whose folding could be obtained in living cells [69,70].

3.4. In-Cell NMR in Insect Cells

The first in-cell NMR study in insect cells was carried out by Hamatsu et al. using GB1, HB8 TTHA1718, rat calmodulin, and human HAH1 as examples [71]. In the study, the target genes were transfected into sf9 cells using a baculovirus system and both ^{15}N-and ^{13}C/^{15}N-labeled proteins were achieved by growing cells in suitable media. In addition to collecting the 2D ^{1}H-^{15}N-HSQC spectrum, the authors collected 3D triple-resonance NMR spectra that are routinely used in backbone assignment (Figure 2). Approximately 80% of signals from backbone atoms were observed, which made the backbone assignment of GB1 possible. The quality of the acquired 3D ^{15}N-seperated NOESY spectrum (Figure 2) was good enough for structural determination as the cross peaks in the spectrum could be assigned [71].

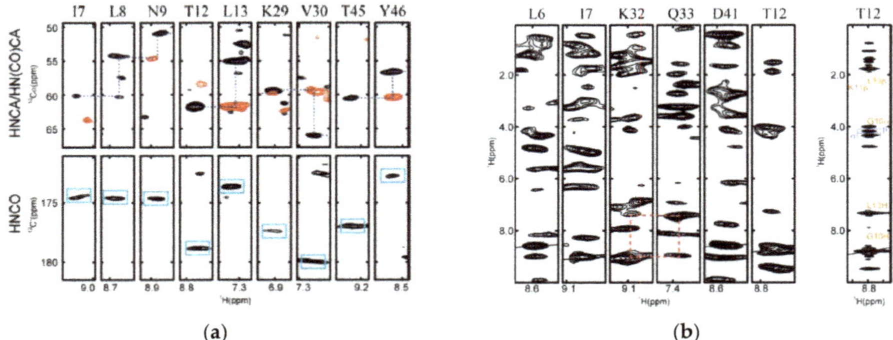

Figure 2. 3D NMR spectra collected in sf9 cells. (**a**) 3D HNCOCA (red), HNCA (black) and HNCO spectra of GB1 in sf9 cells. Blue box indicate the C′ signals. (**b**) Selected strip plot at ^{1}HN$-^1$H dimensions from the 3D a ^{15}N-separated NOESY spectrum of GB1 expressed in sf9 cells. The Cα connectivity and sequential NOEs are indicated as blue and red lines, respectively. This figure was reprinted with permission from the reference [71]. Copyright (2013) American Chemical Society. For more experimental details, please refer to the original publication.

3.5. In-Cell NMR in Human Cells

Overexpression and purification of isotopically labeled proteins from mammalian cells for in vitro NMR studies is normally more challenging than in bacteria due to the experimental cost. In-cell NMR in mammalian cells is important for structural studies of mammalian proteins. To carry out in-cell NMR in mammalian cells, researchers have developed different approaches. One outstanding method is to transform isotopically labeled proteins into the cells through a cell-penetrating peptide (CPP), which is derived from HIV-1 tat protein and can be linked with the target protein through fusion or crossing reactions by disulfide bonds. The structures of ubiquitin and FKBP12 were investigated using this approach [93]. There are several types of CPPs that can be used for protein delivery while the conditions need to be explored in the experiments.

In addition to CPP, toxins were used for delivering isotopically labeled proteins into human cells for in-cell NMR studies. Treatment of nonadherent 293F cells with bacterial toxin streptolysin O (SLO) enabled pore formation on the cell membrane. As the diameter of the pores could reach 35 nm, proteins could reach inside of the cells [91]. Supplying Ca^{2+} in the medium was able to prevent cell death caused by pore formation on the cell membrane to reduce releasing of isotopically labeled protein into the medium [91]. Proteins such as isotopically labeled Tβ4 were able to be delivered into the human cells and exhibited dispersed cross peaks in the NMR spectra [91]. Labeled proteins could also be delivered into cells by electroporation, which was originally used for nucleic acids transfection. Modification of Parkinson's disease protein alpha-synuclein was monitored using in-cell NMR [94,95]. Fusing the target protein with a suitable sequence can localize the protein to desired cellular compartment [67,129], which makes it possible to monitor protein structures in the natural compartments.

In-cell NMR studies in human cells are also achievable using cells with overexpressed proteins. The existing strategies for protein expression in mammalian cells are suitable for producing isotopically labeled proteins for NMR studies [107]. A detail protocol has been developed to produce proteins in mammalian cell lines such as human embryonic kidney 293T (HEK293T) for in-cell NMR studies. In this method, the gene encoding for the target protein is induced into the cells using transient DNA transfection. Isotope-enriched protein is then produced by growing the cells in a medium with ^{15}N-nitrogen sources [116]. This method has advantages over protein delivery, as the target protein is produced directly into the living cells, without any protein purification procedures [107]. Using human SOD1 as an example, the metal binding and effect of copper binding on the redox state of the protein were investigated in the living cells [75]. Folding of Mia40 controlled by cytoplasmic glutaredoxin 1 and thioredoxin 1 was evaluated using in-cell NMR [99]. Mia40 was shown to be stable in the cytoplasm. Such studies provide a view of protein folding in living cells at an atomic level, which is challenging to investigate using other biophysical methods [99]. As the expressed proteins can be translocated to certain sub-cellular compartments, protein structure and folding at certain organelles can be evaluated using solution NMR spectroscopy. Folding of Mia40 and hSOD1 was studied on the intact mitochondria using solution NMR spectroscopy. In addition to proteins, the folding of DNA can also be studied using in-cell NMR. The structure of the DNA i-motif was observed in Hela cells using NMR [101] and the obtained information is useful for future biosensor development.

4. In-Cell NMR in Probing Protein–Protein Interactions

In-cell NMR provides an ideal system to probe protein–protein interactions, as proteins do not exist as a single molecule under the physiological conditions [130]. To probe protein–protein interactions in bacterial cells, the target protein is normally first overexpressed in a M9 medium to achieve isotope labeling (^{15}N). Then the cells were transferred to a normal medium. The binding partner is sequentially induced using another inducer to achieve overexpression. With the extension of induction time, the amount of the binding partner is increased, which is similar to the titration experiment in vitro [131]. Using such a sequential protein expression system, in-cell NMR was used to probe protein–protein interactions in *E. coli* Rosetta (DE3) cells [80]. This study was used to probe the interaction between ubiquitin and proteins with ubiquitin interacting motif (UIM), namely ataxin 3 protein (AUIM) and the signal-transducing adaptor molecule STAM2 [80]. This study provides a unique view of protein–protein interactions in live cells [80].

The number of amino acids that are involved in the molecule interactions might not be correctly estimated in in-cell NMR experiments, as signal broadening is also associated with the formation of stoichiometric complexes in the living cells. To overcome the shortcoming brought about by conventional analysis of the data, Single Value Decomposition (SVD) was proposed to analyze the in-cell NMR binding data [131]. SVD is a mathematical method that can be used to identify the principal components from an arbitrary matrix that was built up from experimental data. SVD has wide applications, and it has been used to process NMR spectra, to determine ligand binding site using information derived from chemical shift perturbations, and to identify allosteric binding sites [132–134]. This method was used to analyze the interactions between the prokaryotic ubiquitin-like protein and mycobacterial proteasome ATPase (Mpa) in living cells [135]. Thioredoxin was shown to have exchanges with other cell components and exhibited a molecular weight of approximately 1 MDa in the living cells. In addition to probing protein–protein interactions, an in-cell NMR study showed that adenylate kinase (ADK) had an open binding pocket binding to ATP and AMP [81]. Human PFN1's specific and unspecific interactions with other proteins were analyzed using in-cell NMR [82]. Accumulated studies have proven that in-cell NMR provides a new avenue to understand protein regulation in the living cells [68].

5. In-Cell NMR in Drug Discovery

In-cell NMR has been shown to be used in different cells, giving rise to the possibility of exploring folding and modification [89] of proteins in physiological environments. Probing protein and drug interactions in living cells is critical in drug discovery, as this information is helpful for medicinal chemists to improve the potency of the compounds. As the interactions are monitored in living cells, it is very helpful to understand the action mode of the developed compounds. Monitoring protein and ligand interactions using in-cell NMR has been successfully carried out in living cells by Banci and Hasnain's team. In their studies, SOD1 was confirmed to form a complex with ebselen, which is an organoselenium compound with broad antioxidant properties [98]. Oxidation of SOD1 in living cells by ebselen was investigated using in-cell NMR. Ebselen was shown to interact with SOD1 and affect its folding in the living cells. This study provides a potential therapeutic application by indicating an unusual SOD1 disulfide bond [98].

5.1. Application of In-Cell NMR in Ligand Screening

Protein and ligand interactions can be demonstrated in living cells by monitoring the signals from the substrate. The enzymatic activity of new Delhi metallo-b-lactamase subclass 1 (NDM-1) expressed in *E. coli* cells can be assayed by monitoring the signals from its substrate meropenem [76]. The inhibition of NDM-1 by inhibitors can be monitored using a ^{1}H-based experiment. This study provides a direct view of the function and inhibition of enzymes in living cells [76]. A similar strategy could also be applied to human cells when the target in drug discovery is from a human being. The NMR spectra of the development compound in the absence and presence of human cells with and without expressed target protein will prove whether the compound binds to the target protein in living cells. Such studies could also be improved to provide more information by incubating compound with human cells harboring different types of target proteins such as mutations. ^{19}F-NMR spectroscopy is also very powerful in in-cell NMR studies, as the background signals from the living cells are reduced because the biological system does not contain fluorine atoms. Cleavage of the fluorinated anandamide analog-ARN1203 was observed in the presence of HEK293 cells harboring expressed fatty acid amide hydrolase (FAAH) [77]. As FAAH is a membrane protein, the assay is feasible using this system, and compound fragments which were able to inhibit its activity were screened and confirmed [136]. The molecular interactions between Bcl-2 and the quercetin-alanine bioconjugate were investigated using proton-based NMR experiments [104]. This study shows that ligand-based NMR such as STD is also applicable in in-cell NMR.

Screening of compounds capable of disrupting protein–protein interactions is feasible using in-cell NMR [84,137]. A system comprising FK506 binding protein 12 (FKBP12) and the 100-residue FKBP-rapamycin binding domain from the mammalian target of rapamycin (FRB) was used in the study. Uniformly ^{15}N-labeled FKBP12 and unlabeled FRB were expressed in *E. coli* using a co-expression system. The complex exhibited a ^{1}H-^{15}N-HSQC spectrum with nicely dispersed cross peaks. Adding rapamycin (binding to FKBP12 with 200 pM affinity) to the solution induced chemical shift perturbations for both FKBP12 and FRB while adding ascomycin to the cell solution induce changes the spectrum of FKBP12 but not FRB, which might be caused by their slightly different binding surfaces on FKBP12. As the existence of two proteins is required to generate the detectable in-cell NMR spectra, this system was then used for screening against a peptide library (Figure 3). Peptides able to disrupt FKBP12 and FRB interactions were identified from a library with 289 dipeptides. The screened peptides were confirmed to disrupt protein–protein interactions in yeast [84] by means of competition experiments with rapamycin and ascomycin. Using a similar method, small molecular compounds that can affect Pup and Mpa interactions were screened from a library consisting of 1597 compounds [83]. To reduce the time for screening, the developed matrix method in which the library compounds were placed a matrix plate and mixed was proven to be a practical and efficient strategy [84].

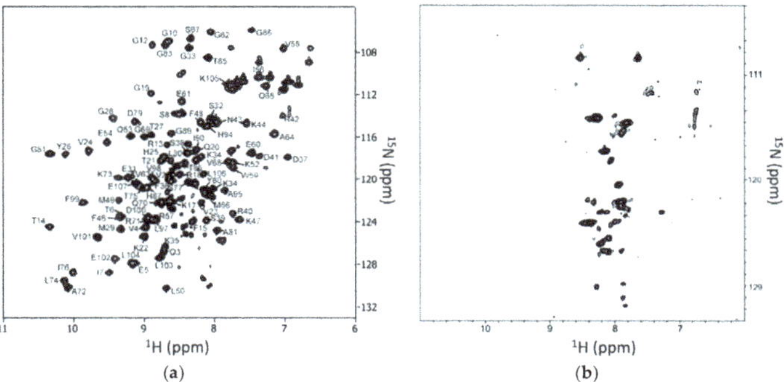

Figure 3. Application of in-cell NMR in compound screening. Peptides that can disrupt FKBP12 and FRB interactions were screened using this approach. (**a**) ^{1}H-^{15}N-HSQC spectrum of FKBP12-FRB complex in which FKBP12 is ^{15}N-labeled. (**b**) Peptide affects the spectrum of FKBP12-FRB complex. This figure was reprinted with permission from the reference [84]. Copyright (2009) American Chemical Society. For more experimental details, please refer to the original publication.

5.2. Application of In-Cell NMR in Target Engagement

Target engagement is a procedure to evaluate protein and ligand interactions in living cells [138]. It is important to understand the molecular action of the developed compounds in drug discovery. As the cellular environment is different from in vitro biochemical environments, the ligand-binding information obtained in vitro might be different from that obtained in vivo. As developed compounds need to be tested in different animal models and different cell lines before they enter into clinical studies, target engagement is therefore critical, as it can provide the real-time binding information in living cells. Several methods have been used in target engagement such as cellular thermal shift assay [139,140] and polarized microscopy [141]. In-cell NMR is a unique tool to study protein–ligand interactions in living cells, suggesting that it can be used as a tool in target engagement [142]. The successful application of in-cell NMR in compound screening and its feasibility for incorporation with other cellular-based mythologies [143,144] make it possible for it to be applied in target engagement.

In-cell NMR was used to validate target engagement of the antituberculosis imidazopyridine amide (IPA) series in living cells [117]. This study used ligand-observed ^{1}H and STDexperiments to confirm drug binding to the cytochrome b in living cells. In addition, the atoms of IPA that are important for interactions were also identified in the binding study, which was helpful for obtaining the structure of the complex. The authors used a heterologous host *M. smegmatis*—a non-pathogenic bacterial system—to avoid the handling of pathogenic bacteria in the NMR spectrometer. This is the first application of an in-cell NMR study in target engagement, and it was encouraging with respect to the possibility of carrying out similar investigations in drug discovery.

6. Perspective

Most structures deposited in the protein data bank are obtained under in vitro conditions, which might differ from those obtained in living cells, as only purified proteins are used in structural determination in vitro. Proteins under national conditions interact with multiple proteins, which cannot be monitored using in vitro structural methods. In-cell NMR will connect the available structures to protein function in vivo. As in-cell NMR studies can be carried out in prokaryotic and eukaryotic cells, cell biology techniques are required to carry out successful in-cell NMR experiments.

In the drug discovery process, probing protein–drug interactions is a critical step in target-based drug discovery. In-cell NMR is therefore a powerful method to evaluate potent compounds in drug development to save experimental cost. The in-cell NMR study in bacteria will be helpful in antibiotic development as both target engagement and compound transportation into the cells can be monitored.

It has been noted that some pathogenic bacteria might not be allowed in NMR studies. The in-cell experiments in mammalian cell lines will be critical both for monitoring protein post modifications and target engagement in developing chemotherapies against human diseases, such as anti-cancer drugs. With the development of new NMR hardware, new methods in sample preparation, and combination with other techniques, in-cell NMR will play more important roles in structural biology and drug discovery.

Funding: This research was funded by Singapore Ministry of Health's National Medical Research Council under its open fund individual research grant (OFIRG17may050, NMRC/OFIRG/0051/2017).

Acknowledgments: The author thanks the funding from National Medical Research Council (NMRC) and support from the Biomedical Research Council (BMRC) of A*STAR.

Conflicts of Interest: The author declares no conflict of interest.

Abbreviations

NMR	Nuclear magnetic resonance
Cryo-EM	cryogenic electron microscopy
STD	Saturation-transfer difference
SOFAST-HMBC	Band-Selective Optimized Flip Angle Short Transient-heteronuclear multiple quantum coherence)
Mia40	Mitochondrial intermembrane space import and assembly protein 40
NOESY	nuclear Overhauser effect spectroscopy)
XT-GB1	SV40 regulatory domain-GB1
Cot17	cytochrome c oxidase copper chaperone
Tβ4	thymosin β4
Bcl-2	B-cell lymphoma 2
GB1	the B domain of G protein
HSQC	heteronuclear single quantum coherence
UIM	ubiquitin interacting motif
hSOD1	human copper, zinc superoxide dismutase 1
SVD	Single Value Decomposition
ADK	adenylate kinase
FBDD	Fragment-based drug discovery
CPP	cell-penetrating peptide
PCS	pseudo-contact shifts
RDC	Residue dipolar coupling
PFN1	human protein profilin 1
FKBP12	FK506 binding protein 12
FRB	the 100-residue FKBP-rapamycin binding domain from the mammalian target of rapamycin
Pup	prokaryotic ubiquitin like protein
Mpa	mycobacterial protease ATPase
HDH	Histidinol Dehydrogensase
GFP	Green Fluorescence Protein

References

1. Billeter, M.; Wagner, G.; Wuthrich, K. Solution NMR structure determination of proteins revisited. *J. Biomol. NMR* **2008**, *42*, 155–158. [CrossRef] [PubMed]
2. Hanske, J.; Sadian, Y.; Muller, C.W. The cryo-EM resolution revolution and transcription complexes. *Curr. Opin. Struct. Biol.* **2018**, *52*, 8–15. [CrossRef] [PubMed]
3. Campbell, I.D. The march of structural biology. *Nat. Rev. Mol. Cell Biol.* **2002**, *3*, 377. [CrossRef] [PubMed]
4. Shi, Y. A Glimpse of Structural Biology through X-Ray Crystallography. *Cell* **2014**, *159*, 995–1014. [CrossRef]

5. Guo, H.; Rubinstein, J.L. Cryo-EM of ATP synthases. *Curr. Opin. Struct. Biol.* **2018**, *52*, 71–79. [CrossRef] [PubMed]

6. Wang, W.; MacKinnon, R. Cryo-EM Structure of the Open Human Ether-a-go-go-Related K(+) Channel hERG. *Cell* **2017**, *169*, 422–430.e10. [CrossRef] [PubMed]

7. Madl, T.; Gabel, F.; Sattler, M. NMR and small-angle scattering-based structural analysis of protein complexes in solution. *J. Struct. Biol.* **2011**, *173*, 472–482. [CrossRef] [PubMed]

8. Zheng, J.; Yong, H.Y.; Panutdaporn, N.; Liu, C.; Tang, K.; Luo, D. High-resolution HDX-MS reveals distinct mechanisms of RNA recognition and activation by RIG-I and MDA5. *Nucleic Acids Res.* **2015**, *43*, 1216–1230. [CrossRef] [PubMed]

9. Iacobucci, C.; Gotze, M.; Ihling, C.H.; Piotrowski, C.; Arlt, C.; Schafer, M.; Hage, C.; Schmidt, R.; Sinz, A. A cross-linking/mass spectrometry workflow based on MS-cleavable cross-linkers and the MeroX software for studying protein structures and protein-protein interactions. *Nat. Protoc.* **2018**, *13*, 2864. [CrossRef] [PubMed]

10. Pielak, G.J.; Li, C.; Miklos, A.C.; Schlesinger, A.P.; Slade, K.M.; Wang, G.F.; Zigoneanu, I.G. Protein nuclear magnetic resonance under physiological conditions. *Biochemistry* **2009**, *48*, 226–234. [CrossRef] [PubMed]

11. Smith, M.J.; Marshall, C.B.; Theillet, F.X.; Binolfi, A.; Selenko, P.; Ikura, M. Real-time NMR monitoring of biological activities in complex physiological environments. *Curr. Opin. Struct. Biol.* **2015**, *32*, 39–47. [CrossRef]

12. Riek, R.; Pervushin, K.; Wuthrich, K. TROSY and CRINEPT: NMR with large molecular and supramolecular structures in solution. *Trends Biochem. Sci.* **2000**, *25*, 462–468. [CrossRef]

13. Hyberts, S.G.; Arthanari, H.; Robson, S.A.; Wagner, G. Perspectives in magnetic resonance: NMR in the post-FFT era. *J. Magn. Reson.* **2014**, *241*, 60–73. [CrossRef]

14. Tugarinov, V.; Choy, W.Y.; Orekhov, V.Y.; Kay, L.E. Solution NMR-derived global fold of a monomeric 82-kDa enzyme. *Proc. Natl. Acad. Sci. USA* **2005**, *102*, 622–627. [CrossRef] [PubMed]

15. Tugarinov, V.; Kay, L.E. An isotope labeling strategy for methyl TROSY spectroscopy. *J. Biomol. NMR* **2004**, *28*, 165–172. [CrossRef] [PubMed]

16. Tugarinov, V.; Kay, L.E. Ile, Leu, and Val methyl assignments of the 723-residue malate synthase G using a new labeling strategy and novel NMR methods. *J. Am. Chem. Soc.* **2003**, *125*, 13868–13878. [CrossRef] [PubMed]

17. Barrett, P.J.; Chen, J.; Cho, M.K.; Kim, J.H.; Lu, Z.; Mathew, S.; Peng, D.; Song, Y.; Van Horn, W.D.; Zhuang, T.; et al. The quiet renaissance of protein nuclear magnetic resonance. *Biochemistry* **2013**, *52*, 1303–1320. [CrossRef]

18. Kay, L.E. New Views of Functionally Dynamic Proteins by Solution NMR Spectroscopy. *J. Mol. Biol.* **2016**, *428 Pt A*, 323–331. [CrossRef]

19. Gayen, S.; Li, Q.; Kang, C. Solution NMR study of the transmembrane domain of single-span membrane proteins: Opportunities and strategies. *Curr. Protein Pept. Sci.* **2012**, *13*, 585–600. [CrossRef]

20. Kang, C.; Li, Q. Solution NMR study of integral membrane proteins. *Curr. Opin. Chem. Biol.* **2011**, *15*, 560–569. [CrossRef] [PubMed]

21. Sanders, C.R.; Sonnichsen, F. Solution NMR of membrane proteins: Practice and challenges. *Magn. Reson. Chem.* **2006**, *44*, S24–S40. [CrossRef] [PubMed]

22. Radoicic, J.; Lu, G.J.; Opella, S.J. NMR structures of membrane proteins in phospholipid bilayers. *Q. Rev. Biophys.* **2014**, *47*, 249–283. [CrossRef] [PubMed]

23. Liang, B.; Tamm, L.K. NMR as a tool to investigate the structure, dynamics and function of membrane proteins. *Nat. Struct. Mol. Biol.* **2016**, *23*, 468–474. [CrossRef]

24. Clark, L.; Dikiy, I.; Rosenbaum, D.M.; Gardner, K.H. On the use of Pichia pastoris for isotopic labeling of human GPCRs for NMR studies. *J. Biomol. NMR* **2018**, *71*, 203–211. [CrossRef] [PubMed]

25. Opella, S.J.; Marassi, F.M. Applications of NMR to membrane proteins. *Arch. Biochem. Biophys.* **2017**, *628*, 92–101. [CrossRef] [PubMed]

26. Oxenoid, K.; Chou, J.J. The present and future of solution NMR in investigating the structure and dynamics of channels and transporters. *Curr. Opin. Struct. Biol.* **2013**, *23*, 547–554. [CrossRef] [PubMed]

27. Shimada, I.; Ueda, T.; Kofuku, Y.; Eddy, M.T.; Wuthrich, K. GPCR drug discovery: Integrating solution NMR data with crystal and cryo-EM structures. *Nat. Rev. Drug Discov.* **2018**. [CrossRef]

28. Hennig, J.; Sattler, M. The dynamic duo: Combining NMR and small angle scattering in structural biology. *Protein Sci.* **2014**, *23*, 669–682. [CrossRef]

29. Li, Y.; Zhang, Z.; Phoo, W.W.; Loh, Y.R.; Li, R.; Yang, H.Y.; Jansson, A.E.; Hill, J.; Keller, T.H.; Nacro, K.; et al. Structural Insights into the Inhibition of Zika Virus NS2B-NS3 Protease by a Small-Molecule Inhibitor. *Structure* **2018**, *26*, 555–564. [CrossRef]

30. Zhang, Z.; Li, Y.; Loh, Y.R.; Phoo, W.W.; Hung, A.W.; Kang, C.; Luo, D. Crystal structure of unlinked NS2B-NS3 protease from Zika virus. *Science* **2016**, *354*, 1597–1600. [CrossRef]

31. Phoo, W.W.; Li, Y.; Zhang, Z.; Lee, M.Y.; Loh, Y.R.; Tan, Y.B.; Ng, E.Y.; Lescar, J.; Kang, C.; Luo, D. Structure of the NS2B-NS3 protease from Zika virus after self-cleavage. *Nat. Commun.* **2016**, *7*, 13410. [CrossRef] [PubMed]

32. Mantsyzov, A.B.; Shen, Y.; Lee, J.H.; Hummer, G.; Bax, A. MERA: A webserver for evaluating backbone torsion angle distributions in dynamic and disordered proteins from NMR data. *J. Biomol. NMR* **2015**, *63*, 85–95. [CrossRef] [PubMed]

33. Shen, Y.; Bax, A. Protein structural information derived from NMR chemical shift with the neural network program TALOS-N. *Methods Mol. Biol.* **2015**, *1260*, 17–32.

34. Shen, Y.; Delaglio, F.; Cornilescu, G.; Bax, A. TALOS+: A hybrid method for predicting protein backbone torsion angles from NMR chemical shifts. *J. Biomol. NMR* **2009**, *44*, 213–223. [CrossRef] [PubMed]

35. Shen, Y.; Bax, A. Homology modeling of larger proteins guided by chemical shifts. *Nat. Methods* **2015**, *12*, 747–750. [CrossRef] [PubMed]

36. Shen, Y.; Bax, A. SPARTA+: A modest improvement in empirical NMR chemical shift prediction by means of an artificial neural network. *J. Biomol. NMR* **2010**, *48*, 13–22. [CrossRef] [PubMed]

37. Zimmerman, D.E.; Kulikowski, C.A.; Huang, Y.; Feng, W.; Tashiro, M.; Shimotakahara, S.; Chien, C.-Y.; Powers, R.; Montelione, G.T. Automated analysis of protein NMR assignments using methods from artificial intelligence11Edited by P. E. Wright. *J. Mol. Biol.* **1997**, *269*, 592–610. [CrossRef] [PubMed]

38. Maciejewski, M.W.; Schuyler, A.D.; Gryk, M.R.; Moraru, I.I.; Romero, P.R.; Ulrich, E.L.; Eghbalnia, H.R.; Livny, M.; Delaglio, F.; Hoch, J.C. NMRbox: A Resource for Biomolecular NMR Computation. *Biophys. J.* **2017**, *112*, 1529–1534. [CrossRef] [PubMed]

39. Lee, W.; Markley, J.L. PINE-SPARKY.2 for automated NMR-based protein structure research. *Bioinformatics* **2018**, *34*, 1586–1588. [CrossRef]

40. Lemke, C.T.; Goudreau, N.; Zhao, S.; Hucke, O.; Thibeault, D.; Llinas-Brunet, M.; White, P.W. Combined X-ray, NMR, and kinetic analyses reveal uncommon binding characteristics of the hepatitis C virus NS3-NS4A protease inhibitor BI 201335. *J. Biol. Chem.* **2011**, *286*, 11434–11443. [CrossRef]

41. Scott, D.E.; Coyne, A.G.; Hudson, S.A.; Abell, C. Fragment-Based Approaches in Drug Discovery and Chemical Biology. *Biochemistry* **2012**, *51*, 4990–5003. [CrossRef] [PubMed]

42. Li, Y.; Liu, S.; Ng, E.Y.; Li, R.; Poulsen, A.; Hill, J.; Pobbati, A.V.; Hung, A.W.; Hong, W.; Keller, T.H.; et al. Structural and ligand-binding analysis of the YAP-binding domain of transcription factor TEAD4. *Biochem. J.* **2018**, *475*, 2043–2055. [CrossRef] [PubMed]

43. Li, Y.; Zhang, Z.; Phoo, W.W.; Loh, Y.R.; Wang, W.; Liu, S.; Chen, M.W.; Hung, A.W.; Keller, T.H.; Luo, D.; et al. Structural Dynamics of Zika Virus NS2B-NS3 Protease Binding to Dipeptide Inhibitors. *Structure* **2017**, *25*, 1242–1250. [CrossRef] [PubMed]

44. Ziarek, J.J.; Baptista, D.; Wagner, G. Recent developments in solution nuclear magnetic resonance (NMR)-based molecular biology. *J. Mol. Med.* **2018**, *96*, 1–8. [CrossRef] [PubMed]

45. Norton, R.S.; Leung, E.W.; Chandrashekaran, I.R.; MacRaild, C.A. Applications of (19)F-NMR in Fragment-Based Drug Discovery. *Molecules* **2016**, *21*, 860. [CrossRef] [PubMed]

46. Gee, C.T.; Arntson, K.E.; Urick, A.K.; Mishra, N.K.; Hawk, L.M.L.; Wisniewski, A.J.; Pomerantz, W.C.K. Protein-observed 19F-NMR for fragment screening, affinity quantification and druggability assessment. *Nat. Protoc.* **2016**, *11*, 1414. [CrossRef]

47. Erlanson, D.A.; Fesik, S.W.; Hubbard, R.E.; Jahnke, W.; Jhoti, H. Twenty years on: The impact of fragments on drug discovery. *Nat. Rev. Drug Discov.* **2016**, *15*, 605–619. [CrossRef]

48. Arntson, K.E.; Pomerantz, W.C.K. Protein-Observed Fluorine NMR: A Bioorthogonal Approach for Small Molecule Discovery. *J. Med. Chem.* **2016**, *59*, 5158–5171. [CrossRef] [PubMed]

49. Yanamala, N.; Dutta, A.; Beck, B.; van Fleet, B.; Hay, K.; Yazbak, A.; Ishima, R.; Doemling, A.; Klein-Seetharaman, J. NMR-based screening of membrane protein ligands. *Chem. Biol. Drug Des.* **2010**, *75*, 237–256. [CrossRef]

50. Hanzawa, H.; Takizawa, T. [NMR screening in fragment-based drug discovery]. *Yakugaku Zasshi* **2010**, *130*, 325–333. [CrossRef] [PubMed]

51. Shuker, S.B.; Hajduk, P.J.; Meadows, R.P.; Fesik, S.W. Discovering high-affinity ligands for proteins: SAR by NMR. *Science* **1996**, *274*, 1531–1534. [CrossRef] [PubMed]

52. Nitsche, C.; Otting, G. NMR studies of ligand binding. *Curr. Opin. Struct. Biol.* **2018**, *48*, 16–22. [CrossRef] [PubMed]

53. Skora, L.; Jahnke, W. 19F-NMR-Based Dual-Site Reporter Assay for the Discovery and Distinction of Catalytic and Allosteric Kinase Inhibitors. *ACS Med. Chem. Lett.* **2017**, *8*, 632–635. [CrossRef] [PubMed]

54. Mello, J.; Gomes, R.A.; Vital-Fujii, D.G.; Ferreira, G.M.; Trossini, G.H.G. Fragment-based drug discovery as alternative strategy to the drug development for neglected diseases. *Chem. Biol. Drug Des.* **2017**, *90*, 1067–1078. [CrossRef] [PubMed]

55. Curtis-Marof, R.; Doko, D.; Rowe, M.L.; Richards, K.L.; Williamson, R.A.; Howard, M.J. 19 F NMR spectroscopy monitors ligand binding to recombinantly fluorine-labelled b′x from human protein disulphide isomerase (hPDI). *Org. Biomol. Chem.* **2014**, *12*, 3808–3812. [CrossRef] [PubMed]

56. Luchinat, E.; Banci, L. In-Cell NMR in Human Cells: Direct Protein Expression Allows Structural Studies of Protein Folding and Maturation. *Acc. Chem. Res.* **2018**, *51*, 1550–1557. [CrossRef] [PubMed]

57. Serber, Z.; Dotsch, V. In-cell NMR spectroscopy. *Biochemistry* **2001**, *40*, 14317–14323. [CrossRef]

58. Lippens, G.; Cahoreau, E.; Millard, P.; Charlier, C.; Lopez, J.; Hanoulle, X.; Portais, J.C. In-cell NMR: From metabolites to macromolecules. *Analyst* **2018**, *143*, 620–629. [CrossRef]

59. Li, C.; Zhao, J.; Cheng, K.; Ge, Y.; Wu, Q.; Ye, Y.; Xu, G.; Zhang, Z.; Zheng, W.; Zhang, X.; et al. Magnetic Resonance Spectroscopy as a Tool for Assessing Macromolecular Structure and Function in Living Cells. *Annu. Rev. Anal. Chem.* **2017**, *10*, 157–182. [CrossRef]

60. Kaplan, M.; Narasimhan, S.; de Heus, C.; Mance, D.; van Doorn, S.; Houben, K.; Popov-Čeleketić, D.; Damman, R.; Katrukha, E.A.; Jain, P.; et al. EGFR Dynamics Change during Activation in Native Membranes as Revealed by NMR. *Cell* **2016**, *167*, 1241–1251.e11. [CrossRef]

61. Renault, M.; Tommassen-van Boxtel, R.; Bos, M.P.; Post, J.A.; Tommassen, J.; Baldus, M. Cellular solid-state nuclear magnetic resonance spectroscopy. *Proc. Natl. Acad. Sci. USA* **2012**, *109*, 4863–4868. [CrossRef] [PubMed]

62. Stadmiller, S.S.; Pielak, G.J. The Expanding Zoo of In-Cell Protein NMR. *Biophys. J.* **2018**, *115*, 1628–1629. [CrossRef] [PubMed]

63. Li, C.; Charlton, L.M.; Lakkavaram, A.; Seagle, C.; Wang, G.; Young, G.B.; Macdonald, J.M.; Pielak, G.J. Differential dynamical effects of macromolecular crowding on an intrinsically disordered protein and a globular protein: Implications for in-cell NMR spectroscopy. *J. Am. Chem. Soc.* **2008**, *130*, 6310–6311. [CrossRef] [PubMed]

64. Warnet, X.L.; Arnold, A.A.; Marcotte, I.; Warschawski, D.E. In-Cell Solid-State NMR: An Emerging Technique for the Study of Biological Membranes. *Biophys. J.* **2015**, *109*, 2461–2466. [CrossRef] [PubMed]

65. Pielak, G.J.; Tian, F. Membrane proteins, magic-angle spinning, and in-cell NMR. *Proc. Natl. Acad. Sci. USA* **2012**, *109*, 4715–4716. [CrossRef] [PubMed]

66. Serber, Z.; Keatinge-Clay, A.T.; Ledwidge, R.; Kelly, A.E.; Miller, S.M.; Dotsch, V. High-resolution macromolecular NMR spectroscopy inside living cells. *J. Am. Chem. Soc.* **2001**, *123*, 2446–2447. [CrossRef] [PubMed]

67. Banci, L.; Bertini, I.; Cefaro, C.; Ciofi-Baffoni, S.; Gallo, A.; Martinelli, M.; Sideris, D.P.; Katrakili, N.; Tokatlidis, K. MIA40 is an oxidoreductase that catalyzes oxidative protein folding in mitochondria. *Nat. Struct. Mol. Biol.* **2009**, *16*, 198–206. [CrossRef]

68. Sakakibara, D.; Sasaki, A.; Ikeya, T.; Hamatsu, J.; Hanashima, T.; Mishima, M.; Yoshimasu, M.; Hayashi, N.; Mikawa, T.; Wälchli, M.; et al. Protein structure determination in living cells by in-cell NMR spectroscopy. *Nature* **2009**, *458*, 102. [CrossRef]

69. Müntener, T.; Häussinger, D.; Selenko, P.; Theillet, F.-X. In-Cell Protein Structures from 2D NMR Experiments. *J. Phys. Chem. Lett.* **2016**, *7*, 2821–2825. [CrossRef]

70. Pan, B.-B.; Yang, F.; Ye, Y.; Wu, Q.; Li, C.; Huber, T.; Su, X.-C. 3D structure determination of a protein in living cells using paramagnetic NMR spectroscopy. *Chem. Commun.* **2016**, *52*, 10237–10240. [CrossRef]

71. Hamatsu, J.; O'Donovan, D.; Tanaka, T.; Shirai, T.; Hourai, Y.; Mikawa, T.; Ikeya, T.; Mishima, M.; Boucher, W.; Smith, B.O.; et al. High-resolution heteronuclear multidimensional NMR of proteins in living insect cells using a baculovirus protein expression system. *J. Am. Chem. Soc.* **2013**, *135*, 1688–1691. [CrossRef] [PubMed]

72. Banci, L.; Barbieri, L.; Bertini, I.; Cantini, F.; Luchinat, E. In-cell NMR in *E. coli* to Monitor Maturation Steps of hSOD1. *PLoS ONE* **2011**, *6*, e23561. [CrossRef] [PubMed]

73. Li, C.; Wang, G.-F.; Wang, Y.; Creager-Allen, R.; Lutz, E.A.; Scronce, H.; Slade, K.M.; Ruf, R.A.S.; Mehl, R.A.; Pielak, G.J. Protein 19F NMR in Escherichia coli. *J. Am. Chem. Soc.* **2010**, *132*, 321–327. [CrossRef]

74. Hough, L.E.; Dutta, K.; Sparks, S.; Temel, D.B.; Kamal, A.; Tetenbaum-Novatt, J.; Rout, M.P.; Cowburn, D. The molecular mechanism of nuclear transport revealed by atomic-scale measurements. *eLife* **2015**, *4*, e10027. [CrossRef] [PubMed]

75. Banci, L.; Barbieri, L.; Bertini, I.; Luchinat, E.; Secci, E.; Zhao, Y.; Aricescu, A.R. Atomic-resolution monitoring of protein maturation in live human cells by NMR. *Nat. Chem. Biol.* **2013**, *9*, 297. [CrossRef] [PubMed]

76. Ma, J.; McLeod, S.; MacCormack, K.; Sriram, S.; Gao, N.; Breeze, A.L.; Hu, J. Real-Time Monitoring of New Delhi Metallo-β-Lactamase Activity in Living Bacterial Cells by ^{1}H NMR Spectroscopy. *Angew. Chem. Int. Ed.* **2014**, *53*, 2130–2133. [CrossRef] [PubMed]

77. Veronesi, M.; Giacomina, F.; Romeo, E.; Castellani, B.; Ottonello, G.; Lambruschini, C.; Garau, G.; Scarpelli, R.; Bandiera, T.; Piomelli, D.; et al. Fluorine nuclear magnetic resonance-based assay in living mammalian cells. *Anal. Biochem.* **2016**, *495*, 52–59. [CrossRef]

78. Serber, Z.; Straub, W.; Corsini, L.; Nomura, A.M.; Shimba, N.; Craik, C.S.; Ortiz de Montellano, P.; Dotsch, V. Methyl groups as probes for proteins and complexes in in-cell NMR experiments. *J. Am. Chem. Soc.* **2004**, *126*, 7119–7125. [CrossRef]

79. Barnes, C.O.; Pielak, G.J. In-cell protein NMR and protein leakage. *Proteins* **2011**, *79*, 347–351. [CrossRef]

80. Burz, D.S.; Dutta, K.; Cowburn, D.; Shekhtman, A. Mapping structural interactions using in-cell NMR spectroscopy (STINT-NMR). *Nat. Methods* **2006**, *3*, 91–93. [CrossRef]

81. Majumder, S.; Xue, J.; DeMott, C.M.; Reverdatto, S.; Burz, D.S.; Shekhtman, A. Probing Protein Quinary Interactions by In-Cell Nuclear Magnetic Resonance Spectroscopy. *Biochemistry* **2015**, *54*, 2727–2738. [CrossRef] [PubMed]

82. Barbieri, L.; Luchinat, E.; Banci, L. Protein interaction patterns in different cellular environments are revealed by in-cell NMR. *Sci. Rep.* **2015**, *5*, 14456. [CrossRef] [PubMed]

83. DeMott, C.M.; Girardin, R.; Cobbert, J.; Reverdatto, S.; Burz, D.S.; McDonough, K.; Shekhtman, A. Potent Inhibitors of Mycobacterium tuberculosis Growth Identified by Using in-Cell NMR-based Screening. *ACS Chem. Biol.* **2018**, *13*, 733–741. [CrossRef]

84. Xie, J.; Thapa, R.; Reverdatto, S.; Burz, D.S.; Shekhtman, A. Screening of Small Molecule Interactor Library by Using In-Cell NMR Spectroscopy (SMILI-NMR). *J. Med. Chem.* **2009**, *52*, 3516–3522. [CrossRef]

85. Banci, L.; Bertini, I.; Cefaro, C.; Cenacchi, L.; Ciofi-Baffoni, S.; Felli, I.C.; Gallo, A.; Gonnelli, L.; Luchinat, E.; Sideris, D.; et al. Molecular chaperone function of Mia40 triggers consecutive induced folding steps of the substrate in mitochondrial protein import. *Proc. Natl. Acad. Sci. USA* **2010**, *107*, 20190–20195. [CrossRef] [PubMed]

86. Sakai, T.; Tochio, H.; Tenno, T.; Ito, Y.; Kokubo, T.; Hiroaki, H.; Shirakawa, M. In-cell NMR spectroscopy of proteins inside Xenopus laevis oocytes. *J. Biomol. NMR* **2006**, *36*, 179–188. [CrossRef]

87. Selenko, P.; Serber, Z.; Gadea, B.; Ruderman, J.; Wagner, G. Quantitative NMR analysis of the protein G B1 domain in Xenopus laevis egg extracts and intact oocytes. *Proc. Natl. Acad. Sci. USA* **2006**, *103*, 11904–11909. [CrossRef]

88. Serber, Z.; Selenko, P.; Hansel, R.; Reckel, S.; Lohr, F.; Ferrell, J.E., Jr.; Wagner, G.; Dotsch, V. Investigating macromolecules inside cultured and injected cells by in-cell NMR spectroscopy. *Nat. Protoc.* **2006**, *1*, 2701–2709. [CrossRef]

89. Selenko, P.; Frueh, D.P.; Elsaesser, S.J.; Haas, W.; Gygi, S.P.; Wagner, G. In situ observation of protein phosphorylation by high-resolution NMR spectroscopy. *Nat. Struct. Mol. Biol.* **2008**, *15*, 321–329. [CrossRef]

90. Bertrand, K.; Reverdatto, S.; Burz, D.S.; Zitomer, R.; Shekhtman, A. Structure of proteins in eukaryotic compartments. *J. Am. Chem. Soc.* **2012**, *134*, 12798–12806. [CrossRef]

91. Ogino, S.; Kubo, S.; Umemoto, R.; Huang, S.; Nishida, N.; Shimada, I. Observation of NMR signals from proteins introduced into living mammalian cells by reversible membrane permeabilization using a pore-forming toxin, streptolysin O. *J. Am. Chem. Soc.* **2009**, *131*, 10834–10835. [CrossRef] [PubMed]

92. Mochizuki, A.; Saso, A.; Zhao, Q.; Kubo, S.; Nishida, N.; Shimada, I. Balanced Regulation of Redox Status of Intracellular Thioredoxin Revealed by in-Cell NMR. *J. Am. Chem. Soc.* **2018**, *140*, 3784–3790. [CrossRef] [PubMed]

93. Inomata, K.; Ohno, A.; Tochio, H.; Isogai, S.; Tenno, T.; Nakase, I.; Takeuchi, T.; Futaki, S.; Ito, Y.; Hiroaki, H.; et al. High-resolution multi-dimensional NMR spectroscopy of proteins in human cells. *Nature* **2009**, *458*, 106–109. [CrossRef] [PubMed]

94. Binolfi, A.; Limatola, A.; Verzini, S.; Kosten, J.; Theillet, F.X.; May Rose, H.; Bekei, B.; Stuiver, M.; van Rossum, M.; Selenko, P. Intracellular repair of oxidation-damaged alpha-synuclein fails to target C-terminal modification sites. *Nat. Commun.* **2016**, *7*, 10251. [CrossRef] [PubMed]

95. Theillet, F.X.; Binolfi, A.; Bekei, B.; Martorana, A.; Rose, H.M.; Stuiver, M.; Verzini, S.; Lorenz, D.; van Rossum, M.; Goldfarb, D.; et al. Structural disorder of monomeric alpha-synuclein persists in mammalian cells. *Nature* **2016**, *530*, 45–50. [CrossRef] [PubMed]

96. Luchinat, E.; Barbieri, L.; Rubino, J.T.; Kozyreva, T.; Cantini, F.; Banci, L. In-cell NMR reveals potential precursor of toxic species from SOD1 fALS mutants. *Nat. Commun.* **2014**, *5*, 5502. [CrossRef] [PubMed]

97. Luchinat, E.; Barbieri, L.; Banci, L. A molecular chaperone activity of CCS restores the maturation of SOD1 fALS mutants. *Sci. Rep.* **2017**, *7*, 17433. [CrossRef] [PubMed]

98. Capper, M.J.; Wright, G.S.A.; Barbieri, L.; Luchinat, E.; Mercatelli, E.; McAlary, L.; Yerbury, J.J.; O'Neill, P.M.; Antonyuk, S.V.; Banci, L.; et al. The cysteine-reactive small molecule ebselen facilitates effective SOD1 maturation. *Nat. Commun.* **2018**, *9*, 1693. [CrossRef]

99. Banci, L.; Barbieri, L.; Luchinat, E.; Secci, E. Visualization of Redox-Controlled Protein Fold in Living Cells. *Chem. Biol.* **2013**, *20*, 747–752. [CrossRef] [PubMed]

100. Mercatelli, E.; Barbieri, L.; Luchinat, E.; Banci, L. Direct structural evidence of protein redox regulation obtained by in-cell NMR. *Biochim. Biophys. Acta (BBA)-Mol. Cell Res.* **2016**, *1863*, 198–204. [CrossRef] [PubMed]

101. Dzatko, S.; Krafcikova, M.; Hänsel-Hertsch, R.; Fessl, T.; Fiala, R.; Loja, T.; Krafcik, D.; Mergny, J.-L.; Foldynova-Trantirkova, S.; Trantirek, L. Evaluation of the Stability of DNA i-Motifs in the Nuclei of Living Mammalian Cells. *Angew. Chem. Int. Ed.* **2018**, *57*, 2165–2169. [CrossRef] [PubMed]

102. Luchinat, E.; Secci, E.; Cencetti, F.; Bruni, P. Sequential protein expression and selective labeling for in-cell NMR in human cells. *Biochim. Biophys. Acta (BBA)-Gen. Subj.* **2016**, *1860*, 527–533. [CrossRef] [PubMed]

103. Barbieri, L.; Luchinat, E.; Banci, L. Intracellular metal binding and redox behavior of human DJ-1. *JBIC J. Biol. Inorg. Chem.* **2018**, *23*, 61–69. [CrossRef] [PubMed]

104. Primikyri, A.; Sayyad, N.; Quilici, G.; Vrettos, E.I.; Lim, K.; Chi, S.-W.; Musco, G.; Gerothanassis, I.P.; Tzakos, A.G. Probing the interaction of a quercetin bioconjugate with Bcl-2 in living human cancer cells with in-cell NMR spectroscopy. *FEBS Lett.* **2018**, *592*, 3367–3379. [CrossRef] [PubMed]

105. Manna, S.; Sarkar, D.; Srivatsan, S.G. A Dual-App Nucleoside Probe Provides Structural Insights into the Human Telomeric Overhang in Live Cells. *J. Am. Chem. Soc.* **2018**, *140*, 12622–12633. [CrossRef] [PubMed]

106. Chatterjee, D.; Zhiping, L.L.; Tan, S.-M.; Bhattacharjya, S. Interaction Analyses of the Integrin β2 Cytoplasmic Tail with the F3 FERM Domain of Talin and 14-3-3ζ Reveal a Ternary Complex with Phosphorylated Tail. *J. Mol. Biol.* **2016**, *428*, 4129–4142. [CrossRef] [PubMed]

107. Luchinat, E.; Banci, L. In-cell NMR: A topical review. *IUCrJ* **2017**, *4 Pt 2*, 108–118. [CrossRef]

108. Pervushin, K.; Riek, R.; Wider, G.; Wuthrich, K. Attenuated T2 relaxation by mutual cancellation of dipole-dipole coupling and chemical shift anisotropy indicates an avenue to NMR structures of very large biological macromolecules in solution. *Proc. Natl. Acad. Sci. USA* **1997**, *94*, 12366–12371. [CrossRef] [PubMed]

109. Riek, R.; Fiaux, J.; Bertelsen, E.B.; Horwich, A.L.; Wuthrich, K. Solution NMR techniques for large molecular and supramolecular structures. *J. Am. Chem. Soc.* **2002**, *124*, 12144–12153. [CrossRef] [PubMed]

110. Cruzeiro-Silva, C.; Albernaz, F.P.; Valente, A.P.; Almeida, F.C. In-Cell NMR spectroscopy: Inhibition of autologous protein expression reduces Escherichia coli lysis. *Cell Biochem. Biophys.* **2006**, *44*, 497–502. [CrossRef]

111. Kubo, S.; Nishida, N.; Udagawa, Y.; Takarada, O.; Ogino, S.; Shimada, I. A Gel-Encapsulated Bioreactor System for NMR Studies of Protein–Protein Interactions in Living Mammalian Cells. *Angew. Chem. Int. Ed.* **2013**, *52*, 1208–1211. [CrossRef] [PubMed]

112. Breindel, L.; DeMott, C.; Burz, D.S.; Shekhtman, A. Real-Time In-Cell Nuclear Magnetic Resonance: Ribosome-Targeted Antibiotics Modulate Quinary Protein Interactions. *Biochemistry* **2018**, *57*, 540–546. [CrossRef] [PubMed]

113. Elowitz, M.B.; Surette, M.G.; Wolf, P.E.; Stock, J.B.; Leibler, S. Protein mobility in the cytoplasm of Escherichia coli. *J. Bacteriol.* **1999**, *181*, 197–203. [PubMed]

114. Robinson, K.E.; Reardon, P.N.; Spicer, L.D. In-cell NMR spectroscopy in Escherichia coli. *Methods Mol. Biol.* **2012**, *831*, 261–277. [PubMed]

115. Keizers, P.H.J.; Ubbink, M. Paramagnetic tagging for protein structure and dynamics analysis. *Prog. Nucl. Magn. Reson. Spectrosc.* **2011**, *58*, 88–96. [CrossRef] [PubMed]

116. Barbieri, L.; Luchinat, E.; Banci, L. Characterization of proteins by in-cell NMR spectroscopy in cultured mammalian cells. *Nat. Protoc.* **2016**, *11*, 1101–1111. [CrossRef] [PubMed]

117. Bouvier, G.; Simenel, C.; Jang, J.; Kalia, N.P.; Choi, I.; Nilges, M.; Pethe, K.; Izadi-Pruneyre, N. Target engagement and binding mode of an anti-tuberculosis drug to its bacterial target deciphered in whole living cells by NMR. *Biochemistry* **2018**. [CrossRef]

118. Tochio, H. Watching protein structure at work in living cells using NMR spectroscopy. *Curr. Opin. Chem. Biol.* **2012**, *16*, 609–613. [CrossRef]

119. Serber, Z.; Ledwidge, R.; Miller, S.M.; Dotsch, V. Evaluation of parameters critical to observing proteins inside living Escherichia coli by in-cell NMR spectroscopy. *J. Am. Chem. Soc.* **2001**, *123*, 8895–8901. [CrossRef]

120. Pastore, A.; Temussi, P.A. The Emperor's new clothes: Myths and truths of in-cell NMR. *Arch. Biochem. Biophys.* **2017**, *628*, 114–122. [CrossRef]

121. Ito, Y.; Mikawa, T.; Smith, B.O. In-cell NMR of intrinsically disordered proteins in prokaryotic cells. *Methods Mol. Biol.* **2012**, *895*, 19–31. [PubMed]

122. Theillet, F.X.; Binolfi, A.; Frembgen-Kesner, T.; Hingorani, K.; Sarkar, M.; Kyne, C.; Li, C.; Crowley, P.B.; Gierasch, L.; Pielak, G.J.; et al. Physicochemical properties of cells and their effects on intrinsically disordered proteins (IDPs). *Chem. Rev.* **2014**, *114*, 6661–6714. [CrossRef] [PubMed]

123. Danielsson, J.; Mu, X.; Lang, L.; Wang, H.; Binolfi, A.; Theillet, F.X.; Bekei, B.; Logan, D.T.; Selenko, P.; Wennerstrom, H.; et al. Thermodynamics of protein destabilization in live cells. *Proc. Natl. Acad. Sci. USA* **2015**, *112*, 12402–12407. [CrossRef] [PubMed]

124. Gibbs, E.B.; Cook, E.C.; Showalter, S.A. Application of NMR to studies of intrinsically disordered proteins. *Arch. Biochem. Biophys.* **2017**, *628*, 57–70. [CrossRef] [PubMed]

125. Majumder, S.; DeMott, C.M.; Reverdatto, S.; Burz, D.S.; Shekhtman, A. Total Cellular RNA Modulates Protein Activity. *Biochemistry* **2016**, *55*, 4568–4573. [CrossRef]

126. Kang, C.; Vanoye, C.G.; Welch, R.C.; Van Horn, W.D.; Sanders, C.R. Functional delivery of a membrane protein into oocyte membranes using bicelles. *Biochemistry* **2010**, *49*, 653–655. [CrossRef]

127. Tian, C.; Vanoye, C.G.; Kang, C.; Welch, R.C.; Kim, H.J.; George, A.L., Jr.; Sanders, C.R. Preparation, functional characterization, and NMR studies of human KCNE1, a voltage-gated potassium channel accessory subunit associated with deafness and long QT syndrome. *Biochemistry* **2007**, *46*, 11459–11472. [CrossRef]

128. Otting, G. Protein NMR Using Paramagnetic Ions. *Annu. Rev. Biophys.* **2010**, *39*, 387–405. [CrossRef]

129. Barbieri, L.; Luchinat, E.; Banci, L. Structural insights of proteins in sub-cellular compartments: In-mitochondria NMR. *Biochim. Biophys. Acta (BBA)-Mol. Cell Res.* **2014**, *1843*, 2492–2496. [CrossRef]

130. Breindel, L.; Burz, D.S.; Shekhtman, A. Interaction proteomics by using in-cell NMR spectroscopy. *J. Proteom.* **2018**, *191*, 202–211. [CrossRef]

131. Burz, D.S.; Shekhtman, A. The STINT-NMR method for studying in-cell protein-protein interactions. *Curr. Protoc. Protein Sci.* **2010**, *61*.

132. Trbovic, N.; Smirnov, S.; Zhang, F.; Bruschweiler, R. Covariance NMR spectroscopy by singular value decomposition. *J. Magn. Reson.* **2004**, *171*, 277–283. [CrossRef] [PubMed]

133. Selvaratnam, R.; Chowdhury, S.; VanSchouwen, B.; Melacini, G. Mapping allostery through the covariance analysis of NMR chemical shifts. *Proc. Natl. Acad. Sci. USA* **2011**, *108*, 6133–6138. [CrossRef] [PubMed]

134. Arai, M.; Ferreon, J.C.; Wright, P.E. Quantitative analysis of multisite protein-ligand interactions by NMR: Binding of intrinsically disordered p53 transactivation subdomains with the TAZ2 domain of CBP. *J. Am. Chem. Soc.* **2012**, *134*, 3792–3803. [CrossRef] [PubMed]

135. Majumder, S.; DeMott, C.M.; Burz, D.S.; Shekhtman, A. Using singular value decomposition to characterize protein-protein interactions by in-cell NMR spectroscopy. *ChemBioChem* **2014**, *15*, 929–933. [CrossRef] [PubMed]

136. Lambruschini, C.; Veronesi, M.; Romeo, E.; Garau, G.; Bandiera, T.; Piomelli, D.; Scarpelli, R.; Dalvit, C. Development of Fragment-Based n-FABS NMR Screening Applied to the Membrane Enzyme FAAH. *ChemBioChem* **2013**, *14*, 1611–1619. [CrossRef]

137. Rahman, S.; Byun, Y.; Hassan, M.I.; Kim, J.; Kumar, V. Towards understanding cellular structure biology: In-cell NMR. *Biochim. Biophys. Acta (BBA)-Proteins Proteom.* **2017**, *1865*, 547–557. [CrossRef] [PubMed]

138. Durham, T.B.; Blanco, M.-J. Target Engagement in Lead Generation. *Bioorganic Med. Chem. Lett.* **2015**, *25*, 998–1008. [CrossRef] [PubMed]

139. Martinez Molina, D.; Jafari, R.; Ignatushchenko, M.; Seki, T.; Larsson, E.A.; Dan, C.; Sreekumar, L.; Cao, Y.; Nordlund, P. Monitoring drug target engagement in cells and tissues using the cellular thermal shift assay. *Science* **2013**, *341*, 84–87. [CrossRef] [PubMed]

140. Jafari, R.; Almqvist, H.; Axelsson, H.; Ignatushchenko, M.; Lundback, T.; Nordlund, P.; Martinez Molina, D. The cellular thermal shift assay for evaluating drug target interactions in cells. *Nat. Protoc.* **2014**, *9*, 2100–2122. [CrossRef]

141. Dubach, J.M.; Kim, E.; Yang, K.; Cuccarese, M.; Giedt, R.J.; Meimetis, L.G.; Vinegoni, C.; Weissleder, R. Quantitating drug-target engagement in single cells in vitro and in vivo. *Nat. Chem. Biol.* **2017**, *13*, 168–173. [CrossRef] [PubMed]

142. Reckel, S.; Lohr, F.; Dotsch, V. In-cell NMR spectroscopy. *ChemBioChem* **2005**, *6*, 1601–1606. [CrossRef] [PubMed]

143. Luchinat, E.; Gianoncelli, A.; Mello, T.; Galli, A.; Banci, L. Combining in-cell NMR and X-ray fluorescence microscopy to reveal the intracellular maturation states of human superoxide dismutase 1. *Chem. Commun.* **2015**, *51*, 584–587. [CrossRef] [PubMed]

144. Mitri, E.; Barbieri, L.; Vaccari, L.; Luchinat, E. 15N isotopic labelling for in-cell protein studies by NMR spectroscopy and single-cell IR synchrotron radiation FTIR microscopy: A correlative study. *Analyst* **2018**, *143*, 1171–1181. [CrossRef] [PubMed]

International Journal of
Molecular Sciences

MDPI

Review

In-Cell NMR: Analysis of Protein–Small Molecule Interactions, Metabolic Processes, and Protein Phosphorylation

Amit Kumar [1,2,3,*], Lars T. Kuhn [1] and Jochen Balbach [2,4,*]

[1] Astbury Centre for Structural Molecular Biology, School of Molecular and Cellular Biology, University of Leeds, Leeds LS2 9JT, UK
[2] Institute of Physics, Biophysics, Martin–Luther–University Halle–Wittenberg, 06120 Halle, Germany
[3] Department of Diabetes, Faculty of Lifesciences and Medicine, King's College London, Great Maze Pond, London SE1 1UL, UK
[4] Centre for Structure und Dynamics of Proteins (MZP), Martin–Luther–University Halle–Wittenberg, 06120 Halle, Germany
* Correspondence: A.Kumar@leeds.ac.uk (A.K.); jochen.balbach@physik.uni-halle.de (J.B.);
Tel.: +44-113-34-37036 (A.K.); +49-345-55-28550 (J.B.)

Received: 30 November 2018; Accepted: 13 January 2019; Published: 17 January 2019

Abstract: Nuclear magnetic resonance (NMR) spectroscopy enables the non-invasive observation of biochemical processes, in living cells, at comparably high spectral and temporal resolution. Preferably, means of increasing the detection limit of this powerful analytical method need to be applied when observing cellular processes under physiological conditions, due to the low sensitivity inherent to the technique. In this review, a brief introduction to in-cell NMR, protein–small molecule interactions, posttranslational phosphorylation, and hyperpolarization NMR methods, used for the study of metabolites in cellulo, are presented. Recent examples of method development in all three fields are conceptually highlighted, and an outlook into future perspectives of this emerging area of NMR research is given.

Keywords: protein NMR; in-cell NMR; in-situ NMR; DNP; review

1. General Introduction to In-Cell NMR

Most biological pathways are controlled by macromolecules. In order to study the structure and function of biomolecules, in vitro studies are usually applied; then, the resulting data are extrapolated to the native cellular environment. Although such approaches provide a wealth of information surrounding the structure–function activity of biomolecules, they lack the context of a native-complex environment [1,2]. The function of molecules in vivo may differ from that determined in vitro because their native network of interactions within the cell is missing.

In-cell NMR spectroscopy provides a direct readout of protein–protein and ligand–protein interactions in the cellular environment. Therefore, this method gains added value to in vitro based techniques such as cryo-electron microscopy, X-ray crystallography, and hydrogen–deuterium exchange mass spectrometry, generating a wealth of information such as changes in structure and/or dynamics between the free and bound forms. However, these in vitro methods may not fully reflect the protein state in vivo, as the experimental conditions and protein constructs are optimized to obtain the best resolution with each respective method. Electron microscopy provides cellular structural features, but physiological temperatures and molecule sizes below 50 kDa are still challenging for high-resolution studies [3]. In-cell NMR spectroscopy is an ideal technique to study (at atomic resolution) the structural features of biomolecules, their function, and their interactions while they remain in their native cellular environment, as reviewed recently (e.g., References [1,2,4,5]).

This technique is non-invasive and provides structural and biochemical details of macromolecules in solution, while applied in living cells over a wide range of parameters including temperature and pH. NMR methods are ensemble methods, meaning that the outcome is sample-averaged information originating from various molecules in many cells. Thus, the information obtained reflects the global properties of molecules without sub-cellular resolution. Historically, in vivo NMR started with studies of small molecules within living organisms/cells. One of the first in-cell NMR approaches to obtain high-resolution information of biomacromolecules (e.g., proteins) was described by Serber et al. [6,7] inside living cells (Figure 1). In their study, the authors chose two globular soluble proteins, the N-terminal domain of bacterial mercuric ion reductase (Nmer) A and human calmodulin, to explore protein in-cell NMR. In their approach, they utilized conventional recombinant protein expression and isotope labeling in bacterial cells, while considering other parameters such as cell growth, induction time, cell viability, isotopic labeling type, and NMR line broadening. Expression was optimized to obtain sufficient signal, much above that of other cellular proteins, to be detected by NMR spectroscopy.

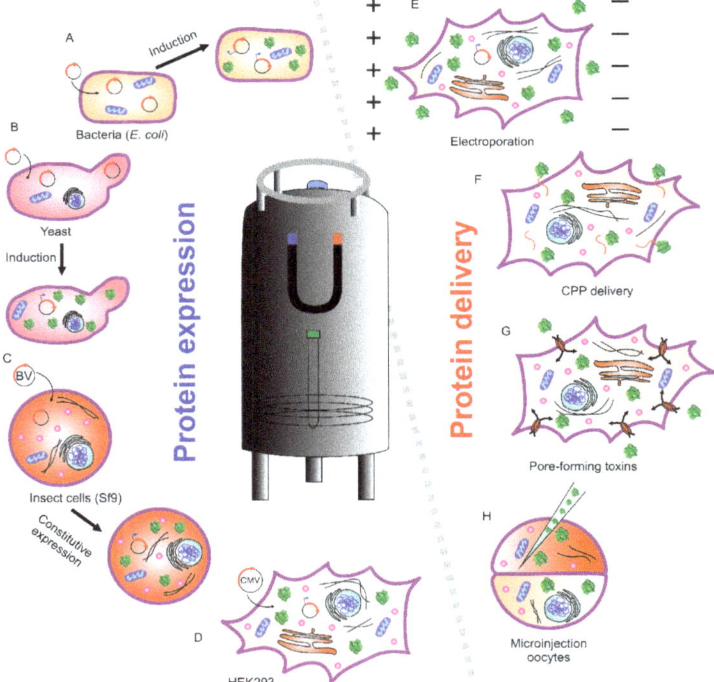

Figure 1. Schematic overview of different known approaches in in-cell NMR. (Left) Endogenously expressed and isotopically labeled protein can be achieved by transferring the expression vector containing the gene of interest into (**A**) bacteria, (**B**) yeast, (**C**) insect cell lines, and (**D**) mammalian cells. (Right) An alternate way of in-cell NMR, where isotopically labeled protein is exogenously prepared followed by delivery into eukaryotic cells with different methods such as (**E**) electroporation, (**F**) attaching protein with cell-penetrating peptides (CPP), (**G**) protein transport via pore-forming toxins, and (**H**) microinjection-mediated delivery into *Xenopus leavis* oocytes.

Different types of isotope labeling of protein samples for in-cell NMR are available. Uniform ^{15}N labeling was found to be most useful and the first choice for most of the studies (Figure 1a,b). The higher natural abundance of ^{13}C in biomolecules, compared to ^{15}N, renders this carbon isotope as the sole modification unsuitable for in-cell NMR studies. An alternative approach to uniform ^{13}C enrichment is the specific labeling of amino acids [7]. Here, methyl-^{13}C methionine labeling was a successful strategy to detect side-chain carbons well above the cellular background [8]. Yet another approach is the incorporation of non-natural amino acids containing ^{19}F. This approach turned out to be a feasible means of investigating protein dynamics in the cellular environment. The advantage of ^{19}F-labeled protein is that the in-cell NMR spectrum is virtually free of background [9,10].

Further developments of in-cell NMR led to methods such as structure interactions NMR (STINT-NMR), cross-correlated relaxation-induced polarization transfer NMR (CRIPT-NMR), and small-molecule interactor libraries NMR (SMILI-NMR). STINT-NMR allowed the study of protein–protein interactions while two molecules are heterologously overexpressed at different time points inside the same bacteria. Firstly, the ^{1}H–^{15}N HSQC spectrum of the ^{15}N-labeled protein of interest is recorded within the cellular environment. Following this, the ^{15}N growth medium is exchanged with an unlabeled medium to overexpress the interaction partner inside the cell. The changes in the chemical environment of the ^{15}N nuclei are observed with time as the concentration of unlabeled binding partner increases. Burz et al. first demonstrated STINT-NMR applications by studying the interaction between a ubiquitin-binding peptide and the signal transducing adaptor molecule 2 protein (STAM2) [11,12]. Subsequently, STINT-NMR was applied to study the interactions between prokaryotic ubiquitin-like protein Pup-GGQ, mycobacterial proteasomal ATPase, Mpa, and the Mtb proteasome core particle (CP). These studies addressed the question of transient binding of Mpa to the proteasome CP that eventually controls the fate of Pup [13]. CRIPT-NMR is yet another in-cell NMR method that allows the identification of interacting surfaces presented on target ^{15}N-labeled proteins within eukaryotic cells, such as HeLa [14]. High-molecular-weight protein molecules can be studied in cells using relaxation optimized ^{15}N-edited cross-relaxation enhanced polarization transfer (CRINEPT), heteronuclear multiple quantum coherence (HMQC), transverse relaxation optimized spectroscopy (TROSY) (^{1}H–^{15}N CRINEPT–HMQC–TROSY) experiments. This method is advantageous due to its relative insensitivity to unavoidable magnetic field inhomogeneity and its high sensitivity to NMR signals. In the in-cell NMR experiment, proton relaxation was minimized by exchanging α and β protons of the amino acids for deuterons called reduced proton density (REDPRO) labeling. Thereafter, a calibration of the CRINEPT transfer time is required to achieve maximum in-cell NMR peak intensities. The in-cell NMR spectrum of the fully expressed protein is compared with its in vitro spectrum and its spectrum in cell lysate. Thus, the interacting surfaces are mapped based upon the residues exhibiting the greatest change in peaks position/intensity. SMILI-NMR was developed, by the same authors, to follow the interactions of proteins with small molecules by in-cell NMR. This technique relies on complex formation of isotope-labeled proteins with small molecules to screen in cellulo entire libraries. The protein of interest gets uniformly labeled with NMR-active heteronuclei under in-cell NMR conditions. This is followed by addition of cell-penetrable small molecules. Monitoring in-cell NMR protein spectra, thus, allows direct observation of protein–small molecule complex formation, in addition to any possible conformational changes [15].

The comprehensive in-cell NMR methods described above to reveal protein–protein or protein–small molecule interactions could potentially act as a bridge between structural and cellular biology. These techniques, already providing excellent results within bacterial systems, unleashed their full potential when applied to eukaryotic and mammalian cell systems. Yeast expression systems provide a simple platform for the study of eukaryotic protein molecules (Figure 1b). This system has the advantage of a unicellular organism with an established expression system and supplement control. The study of proteins within different cellular compartments can be readily performed in yeast [16]. Although the yeast expression system is quite valuable, it suffers from the short lifetime of cells in the NMR sample tube, limiting the experimental observation of events to just a few hours. To overcome this limitation, micro-bioreactors are available for both bacteria/yeast and human cells, which can supply fresh medium and air, and maintain a stable pH value [17,18].

In-cell NMR was first performed in eukaryotic cells on the *Xenopus laevis* oocyte cell system (Figure 1H) [19–21]. This was achieved by preparing protein, injecting into the oocytes, resulting in high labeling selectivity and almost no cellular background. This method proved to be an excellent tool to study posttranslational protein modifications. The Selenko group and others studied serine, threonine, or tyrosine phosphorylation in physiological environments using reconstituted kinase reactions, cell extracts, and intact cells [22]. Real-time monitoring provides additional information about mechanistic insights into modification hierarchies. These may include inhibitory [23], sequential [24], stimulatory [25,26], or "priming" events in phosphorylation cascades. The stepwise modifications of adjacent casein kinase 2 binding sites in the SV40 large T antigen regulatory region, coupled to intermediate substrate release, was disclosed using in situ NMR within *Xenopus laevis* egg extracts and whole live oocyte cells [26].

Inomata et al. applied in-cell NMR spectroscopy in cultured human cells utilizing the cell-penetrating peptide (CPP)2 derived from the human immunodeficiency virus (HIV-1) trans-activator of transcription (Tat) protein. The method relies upon the fusion of (CPP)2 to the protein of interest for internalization into cells (Figure 1F). It is also possible to covalently link (CPP)2 to the protein of interest via a disulfide bond. Such a linkage is cleaved upon internalization in the reducing cellular environment, releasing the peptide-free protein [27]. Using this approach, the folding of human superoxide dismutase (SOD1) during individual steps of the maturation process was studied [28,29]. Its misfolding is implicated in Lou Gehrig's disease, leading to fatal motor neuron impairments. An alternate approach was provided by Ogino et al., who used pore-forming toxins (streptolysin O) to permeabilize the plasma membrane, followed by resealing the plasma membrane with Ca^{2+} to prevent cell death. This allows a sufficient amount of labeled protein to translocate into cultured human cells (Figure 1G). Thymosin β4 (Tβ4) was delivered via this method to 293F cells. The authors observed N-terminal acetylation of Tβ4, which occurred inside the cell as a posttranslational modification [30].

Electroporation is an efficient and recently employed method to transfer isotope-labeled protein into mammalian cells [5] The reversible permeabilization of the plasma membrane allows protein internalization via passive diffusion (Figure 1E). The Banci group expressed protein intracellularly within cultured human cells (Figure 1D) [31]. With this approach, it is possible to obtain atomic-resolution information pertaining to protein folding and maturation processes occurring immediately after protein synthesis in the cytoplasm, using NMR spectroscopy [32,33]. The expression and subsequent NMR analysis of proteins in insect cells was also reported (Figure 1C). For example, in-cell NMR spectra could be recorded using the sf9 cell/baculovirus system for four small model proteins (*Streptococcus* protein G B1 domain, *Thermus thermophilus* HB8 TTHA1718, rat calmodulin, and human HAH1 [34]). Electroporation was also successful in studying the phosphorylation pattern of the intrinsically disordered tau protein by in-cell NMR [35]. Here, disease-associated phosphorylation was immediately eliminated after delivery into human embryonic kidney (HEK)-293T cells. Further examples of in-cell NMR studies of protein phosphorylation are discussed in detail below.

More recently, interactions between unlabeled proteins and small molecules became accessible under in-cell NMR conditions. Here, for the first time, the interaction of unlabeled anti-apoptotic protein B-cell lymphoma 2 (Bcl-2) with the quercetin–alanine bioconjugate was studied in living human cancer cells utilizing saturation transfer difference (STD) and transfer NOESY (Tr-NOESY) NMR experiments [36].

2. In-Cell NMR and Small Molecules

In recent years, in-cell NMR was utilized for probing protein structures, protein folding, disulfide-bond formation, protein–protein and protein–small molecule interactions, and metal uptake in living cells [27,37–40]. In general, it is quite challenging to probe protein–protein and protein–small molecule interactions within living cells. Difficulties surrounding this include poor spectral quality caused by specific and non-specific interactions. However, recent developments showed the success of in-cell NMR in probing protein–small molecule interactions.

2.1. Protein–Small Molecule Interactions

SMILI-NMR is an exciting example [15]. Once the target protein shows detectable and well-dispersed cross-peaks, NMR can be used to carry out target-engagement drug discovery inside the cell. A second example is the interaction of 12-kDa FK506-binding protein (FKBP12) within living cells with extracellularly administered immuno-suppressants [41]. Here, a cleavable CPP–ubiquitin–FKBP12 construct was prepared and the ^{15}N-labeled fusion protein was expressed and purified from *Escherichia coli*. The labeled protein was transduced into HeLa cells. Following this, the CPP–ubiquitin was cleaved off by endogenous deubiquitinating enzymes of the HeLa cells. Subsequently, CPP aggregated, leaving ^{15}N FKBP12 as the sole soluble, labeled protein in the cytosol. Its cross-peak pattern was similar to ^{15}N FKBP12 measured in vitro. This indicated that, in HeLa cells, the three-dimensional structure of FKBP12 was maintained and, therefore, the interaction with small molecules could be studied. Subsequently, in-cell NMR spectra were recorded after treatment with the immune-suppressants FK506 or rapamycin. Significant spectral changes were observed after administration of these drug molecules. Interestingly, the two spectra recorded in vitro and in cellulo had both similar features and subtle changes. Thus, some interactions may be correlated with the interaction observed only in living cultured cells [27,41].

In contrast to classical in-cell NMR experiments, in many cases, the protein cannot be labeled externally and then transduced into the cells of interest. Instead, two steps of production and isotopic labeling need to be carried out simultaneously. Special medium and labeling protocols are well established for bacteria and other organisms, including yeast and insect cells. The use of bacteria to overexpress protein within cells may lead to heavy background signals, caused by the concomitant labeling of other cellular components. To tackle this problem, the Dötsch group developed a scheme which reduced the background noise observed when studying the bacterial protein Nmer A. The activity of bacterial RNA polymerase can be inhibited by rifampicin but not by bacteriophage T7. Since protein expression was under the control of the T7 promoter, production of all endogenous bacterial proteins could be suppressed by rifampicin [6,7]. Nmer A plays a critical role in the bacterial pathway involved with mercury detoxification. The addition of Zn^{2+} to the NMR tube led to changes/disappearance of cross-peaks, thus highlighting the possible utility of this approach to investigate metal and drug binding.

Recently, we showed the targeting of the bacterial chaperone "sensitive to lysis" (SlyD) to inhibit bacterial growth using a small molecule, with in-cell NMR spectroscopy [4]. Emergence of dangerous multi-drug-resistant strains of bacteria is one of the biggest threats to human health currently. With a continuous rise in antibacterial resistance, it was estimated that, by 2050, it may result in the death of 10 million people per year [42]. Among *Enterococcus faecium*, *Staphylococcus aureus*, *Klebsiella pneumoniae*, *Acinetobacter baumannii*, *Pseudomonas aeruginosa*, and *Enterobacter* species (ESKAPE) pathogens, Gram-negative bacteria are of particular concern, due to their increased ability to attain multi-drug resistance [43]. In the case of Gram-negative bacteria, many small molecules including antibiotics become ineffective, due to the presence of an outer polysaccharide layer and multi-drug efflux transporters. This led researchers to create a distinct class of novel antibacterial agents. For example, the heat shock proteins (HSPs) were targeted as potential molecules in cancer therapy [44,45]. Much attention was paid to target HSP90, leading to the identification of geldanamycin and radicicol; however, HSP60, HSP70, or other chaperones are less studied so far [46]. Therefore, we chose a small molecule, which was a metal-based coordination complex with a water-soluble organic moiety capable of crossing the cell-wall barrier and selectively targeting the bacterial chaperone SlyD.

SlyD is a bacterial chaperone, and all prokaryotes and archaea express homologous proteins [47,48]. Thus, being unique to prokaryotes and archaea, SlyD represents a potential target against which to develop drug molecules. In these organisms, SlyD is involved in several biochemical pathways including the biosynthesis of [NiFe] hydrogenases, twin-arginine-mediated translocation (Tat transport), and metal storage/release. Additionally, SlyD exhibits both a peptidyl–prolylisomerase (PPIase) and chaperone activity, which prevents protein aggregation by binding to hydrophobic patches [47–49]. The PPIase and molecular chaperone activities [50] are located on two separate domains, whose cooperative interplay is required for full enzymatic activity [47,51–54]. The N-terminal tail contains the Ni^{2+} binding site followed by the FKBP binding domain. The FKBP domain harbors the active site of the PPIase, which is modulated by Ni^{2+}, and the chaperone function is located on a domain that is inserted into the FKBP domain (Figure 2) [47,51–54].

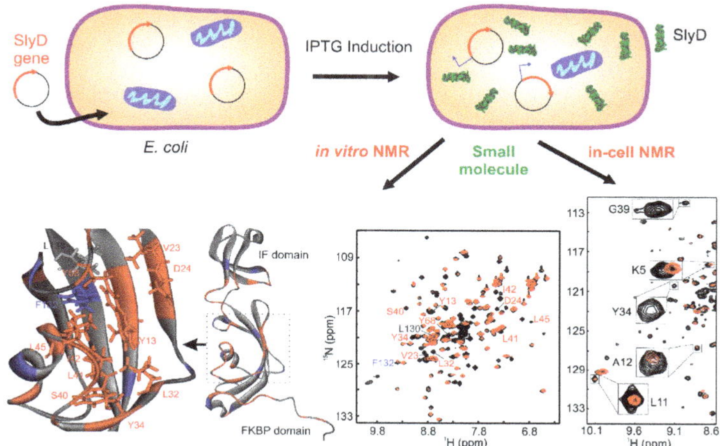

Figure 2. In-cell NMR study of protein–small molecule interactions. Here, a "sensitive to lysis" (SlyD)-containing plasmid was transformed into *Escherichia coli* and protein expression was initiated by isopropyl β-D-1-thiogalactopyranoside (IPTG) induction. These cells were further incubated with a small molecule (Cu^{2+} complex). The interaction of this small molecule with SlyD could be observed using in-cell NMR, and correlated with corresponding in vitro NMR studies, revealing the binding site in SlyD for this Cu^{2+} complex (adopted according to Reference [4]).

Complete bacterial growth can be inhibited with a small molecule (anthracenyl terpyridine Cu^{2+} complex) at 2 μM [4]. In order to evaluate the role of SlyD in bacterial growth, we transformed *E. coli* with the plasmid containing the *SlyD* gene, which is under the control of T7 promoter. The expression of the *SlyD* gene can, thus, be initiated by isopropyl β-D-1-thiogalactopyranoside (IPTG). Overexpressed *SlyD* could restore bacterial growth, confirming that SlyD is involved in growth inhibition induced by the Cu^{2+} complex. This small molecule binds to SlyD with a dissociation constant (K_D) ~50 μM with a 2:1 stoichiometry without inducing large conformational changes. In vitro NMR with ^{15}N SlyD showed that this Cu^{2+} complex binds into the PPIase active site of the FKBP domain (Figure 2), as shown in earlier studies [54]. In-cell NMR spectroscopy confirmed with residue resolution that the Cu^{2+} complex also binds to SlyD inside bacteria, verifying that the complex can penetrate the cell wall and bind to SlyD when inhibiting cell growth. Interestingly, the small molecule also inhibited the growth of pathogenic bacteria from the category of ESKAPE pathogens. The half maximal inhibitory concentration (IC_{50}) of the small molecule was below 1 μM for *Staphylococcus aureus* and *Pseudomonas aeruginosa* [4,55], showing the general applicability of the approach.

2.2. Small-Molecule Libraries

The Shekhtman group developed small-molecule interactor libraries NMR (SMILI-NMR). The interactions between two or more components of a biomolecular complex may be disrupted or enhanced by a few of these small molecules. The method provides atomic-level information and relies on the formation of a defined complex under in-cell NMR conditions. The approach was applied to the fujimycin (FK-506) binding protein (FKBP) and the FKBP rapamycin-binding domain of the mammalian target of rapamycin (FRB), a well-studied system for heterodimer formation, to screen those small molecules that can facilitate hetero-dimerization. In mammalian cells, one of the immune-modulatory systems (mitogenic responses) is constituted by rapamycin–FKBP–FRB interactions [56]. The labeled proteins were sequentially overexpressed using two compatible plasmids in *E. coli* under in-cell NMR conditions. To observe the NMR spectrum of FKBP or FRB, bi-complex (FKBP–FRB) formation was required. The X-ray structure of the FKBP–rapamycin–FRB ternary complex indicated the limited availability of the FKBP–FRB interaction surface. While the interaction between free FKBP and FRB is quite weak, with K_D values >50 μM, a well-defined ternary complex is formed when the FKBP–rapamycin complex binds to FRB (K_D ~12 nM). In their in-cell NMR experiments, either FKBP or FRB was labeled when the complex was formed inside the cells. This did not result in any visible spectrum. Addition of rapamycin to the cell suspension resulted in the appearance of FKBP resonances and, for 32 out of the 107 residues, changes with respect to free FKBP could be detected. Many resonances in FRB residues also changed upon performing the reverse experiment [15]. Ascomycin is a competitive inhibitor of rapamycin. It binds to FKBP with 1.4 nM affinity and has no known affinity with FRB. However, the ascomycin–FKBP complex binds to FRB with lower affinity [57]. Addition of ascomycin to the aforementioned dual-plasmid system resulted in the appearance of the FKBP/FRB NMR spectrum. In the next step, the authors screened a library of small molecules of 289 dipeptides with SMILI-NMR. The peptides were selected as drug candidates from the literature, based on their facile and cost-effective preparation, as well as their ability to be imported into prokaryotic and eukaryotic organisms through naturally occurring transport systems. Screening of a matrix of 17 × 17 peptides resulted in the identification of various combinations that showed completely different behavior when compared to the rapamycin-induced ternary complex. Addition of these dipeptides resulted in extreme line broadening and the disappearance of some peaks in the NMR spectrum. Later on, the addition of low concentrations of Ala–Glu resulted in the same interactions with FKBP, suggesting that FKBP and FRB hetero-oligomerization can be facilitated by this peptide. Thus, SMILI-NMR could screen the protein–small molecule interactions within the cellular environment using high-resolution NMR as a readout [15].

3. In-Cell NMR Observation of Metabolic Processes Using Hyperpolarization

NMR spectroscopy provides considerable opportunities for collecting diverse and unique information on cellular processes due to its non-invasive nature [58], rendering it highly suitable for studying the fate of metabolites and biochemical pathways in situ. In order to raise the detection limit of this inherently insensitive technique, hyperpolarization NMR methods—mainly dissolution dynamic nuclear polarization (d-DNP) and *para*-hydrogen-induced polarization (PHIP)—were devised as tracer techniques to follow the fate of detectable, small molecular probes for the visualization of cellular functions that are not easily observable by other means. Furthermore, NMR signals from these molecules can be enhanced by several orders of magnitude via the combined use of hyperpolarization (HP) and isotope enrichment. The NMR signal enhancements achieved using these methods are often sufficiently high to track endogenous molecules at physiological concentrations. Among all the potentially suitable target molecules for these specialized experiments, hyperpolarized pyruvate, a metabolite whose cell biochemistry lies at the interface between catabolic and anabolic metabolism, is the most widely studied probe. This is mainly due to its (a) high hyperpolarizability, (b) rapid cellular uptake, and (c) central biochemical position as a key intermediate in several biochemical pathways. In addition, a host of other probes emerged in the meantime that can also be used to characterize the phenotype of cells under a particular set of conditions (Table 1).

Table 1. Selection of endogenous target molecules used as reporter probes for hyperpolarization-assisted in-cell NMR studies of metabolic processes, together with their associated field(s) of application.

Target Molecule/Probe	Field of Application (FOA)/Observable
Pyruvate	Pyruvate metabolism; cell permeability; cell lysis; drug efficacy; enzyme activity and reaction fluxes; intracellular pH determination; oncogene signaling; indication of aerobic glycolysis; tricarboxylic acid (TCA) pathway activity; mono carboxylate transporter level/activity; tumor grading
Fumarate	Fumarate metabolism; cell permeability; cell lysis; drug efficacy
Lactate	Enzyme activity and reaction fluxes; tumor grading
Alanine	Enzyme activity and reaction fluxes; enzyme mechanistic studies; tumor grading
Glucose	Gene expression/loss; glycolysis pathway activity; sulfite cytotoxicity; glucose transporter level/activity
Acetate	Enzyme activity and reaction fluxes; intracellular pH determination
Glutamine	Enzyme activity and reaction fluxes
Fructose	Enzyme mechanistic studies

Experimentally, the method, which can lead to a nuclear sensitivity enhancement of up to five orders of magnitude, works as follows: a frozen solution (T ~1.1 to 1.5 K) of the sample to be analyzed is polarized in the presence of a radical molecule, e.g., trityl (triphenylmethyl, "Trityl OX063") or the nitroxide-based radical 2,2,6,6-tetramethylpiperidin-1-yl)oxyl (TEMPO) [59], with microwave (MW) irradiation (spin polarization is transferred via the DNP mechanism from electrons to nuclei upon microwave irradiation at or near the Larmor frequency of the radical electron [60,61]) using a specifically designed DNP polarizer. Subsequently, the sample is thawed, dissolved in a suitable hot solvent, and then transferred to a conventional liquid-state NMR spectrometer for detection [62]. In solids, DNP is known to occur via a number of different mechanisms known as the solid effect, thermal mixing, and the cross effect [60]. Depending on the experimental conditions used in each d-DNP experiment, e.g., radical type, substrate, and solvent (among others), the contribution of each of these effects to the observed polarization enhancement differs.

The longitudinal relaxation time (T_1) of the hyperpolarized nuclei is crucial in dissolution DNP. To preserve the nuclear polarization acquired in the solid state, polarized samples need to be thawed and transferred to the NMR spectrometer faster than nuclear T_1 spin–lattice relaxation. Typical dissolution DNP samples experience a so-called "transfer time" as they are moved from the polarizing magnet to the NMR spectrometer. During this period, the sample is often exposed to low magnetic fields, which are typically on the order of ca. 0.5 mT. Given that nuclear T_1 times are generally shorter at low magnetic fields—and even more so in the presence of radicals—a fast transfer of the sample, as well as the elimination of stable free radicals in solution, is crucial to reduce polarization losses [63]. The use of a "magnetic tunnel" for transfer and/or the addition of a radical scavenging agent, e.g., ascorbate (vitamin C), to the sample during the dissolution step were shown to alleviate polarization losses during and immediately after the transfer. Attempts to shorten the transfer time include construction of a dedicated "hybrid" spectrometer, comprising a single dual-isocenter superconducting magnet featuring regions of different magnetic field strengths suitable for both polarization and NMR detection [64] or, alternatively, a "shuttle" DNP spectrometer comprising a two-center magnet [65].

A significant number of d-DNP-based investigations were carried out in recent years to probe the metabolic behavior of tissues, e.g., heart, liver, and tumor cells, and to study the fate of individual metabolites both in vitro and in vivo [66]. In addition, metabolite molecules hyperpolarized using the d-DNP method were used to probe a variety of different biochemical pathways. For example, the enzymatic conversion of pyruvate to lactate, acetylcarnitine, citrate, and glutamate was tracked in real time employing [2-^{13}C] pyruvate, in isolated perfused heart tissue, to study healthy and pathological states [67]. Hyperpolarized probes were also used to track intracellular pathways of short-chain fatty acids and ketone body metabolism in real time. A butyrate probe visualized the flux of fatty acids to acetoacetate and several tricarboxylic-acid-cycle intermediates in cardiac muscle cells [68]. In addition, hyperpolarized [1-^{13}C] pyruvate was used as a clinical diagnostic tool in metabolic imaging to characterize differences between healthy tissue and tumor cells [69]. Further applications include the study of enzyme kinetics [70], biosynthetic pathways [70], and the detection of lowly populated reaction intermediates [71]. More recently, the *para*-hydrogen-induced hyperpolarization method was also introduced for the signal amplification of metabolites and the subsequent tracking of their biochemical pathways in both healthy and pathological forms of tissue. Up to this point, however, only a few very recent examples exist in the literature where the method was successfully employed to hyperpolarize metabolites for in-cell NMR studies [72–74].

In the following subsections, we give an overview of hyperpolarization NMR methods—in particular, dissolution dynamic nuclear polarization (d-DNP) and *para*-hydrogen-induced nuclear polarization (PHIP)—and a few selected examples of hyperpolarization-assisted in-cell NMR observations of metabolic processes are highlighted in a conceptual manner. The section concludes with an outlook into future perspectives of this emerging, yet still relatively novel, area of NMR research.

3.1. Dissolution DNP Application to Metabolic Pathways and Biological Functionality

The number of applications of heteronuclear d-DNP for the study of living cellular systems is vast [66]. For example, the real-time tracking of metabolic conversion using hyperpolarized NMR is particularly suitable for the observation of metabolic reaction networks, provided that conversion rates are high and that the obtained levels of hyperpolarization are significant. In this context, glycolysis was identified as an adequate metabolic process to be studied with d-DNP, given its overwhelming biochemical importance and its central role in a variety of different biochemical reaction routes. For example, enzymatic reaction mechanisms, bottlenecks, and off-pathway reactions were probed using hyperpolarized carbohydrates, i.e., [2-^{13}C] fructose and [U-^{13}C, U-^{2}H] glucose, as substrates [66]. Chemical detail in the observation of pathway reactions extends to the distinction of isomers and their susceptibility to enzymatic turnover. The use of site-specifically labeled [2-^{13}C] fructose, for example, permitted the real-time observation of probe flux during gluconeogenesis, as well as the formation of non-productive off-pathway intermediates, such as dihydroxy acetone phosphate hydrate [75]. A recent approach combined hyperpolarized dynamic measurements with metabolite extraction, isotopomer evaluation, and flow analysis [76]. This approach measured pyruvate metabolism in living cells to obtain quantitative data of several biochemical pyruvate pathways in different cell types.

The existence of different interlocked pathways, all featuring a similar set of key metabolites, makes it difficult to predict how cellular physiology and intracellular metabolism respond to the modification of individual genes [77]. The use of hyperpolarized NMR spectroscopy to study genetically well-defined and homogeneous cell suspensions shows promise in studying the cellular response to genetic modifications. For example, the two *E. coli* strains, BL21 and K-12, show strong differences in the reaction progression of their pentose phosphate pathways. In the BL21 strain, a reactive intermediate accumulates, and is responsible for covalent modifications observed for the recombinant proteins expressed within this strain [78]. Genome alignment techniques prove that the gene encoding for lactonase—the enzyme which catalyzes the hydrolysis of 6-phosphogluconolactone—is absent in the BL21 strain due to a deletion. Such molecular phenotypes can be observed in the absence of phenotypic variations [79]. Metabolic differences in different cell types were recently compared in human cells by tracking the glycolytic pathway [80]. Ratiometric measurements of lactate and pyruvate signals in two different proliferating cell types were used to non-invasively detect differences in the cytosolic redox state. In the cytosol, lactate and pyruvate form a redox pair, whose equilibration rate depends crucially on the ratio of oxidized to reduced nicotinamide adenine dinucleotide (NAD$^+$/NADH) in the cytosol. PC3, a specific prostate cancer cell line, showed a fourfold increase in the intracellular ratio of free cytosolic NAD$^+$/NADH in comparison with breast cancer cells in an experiment that used hyperpolarized glucose as a reporter metabolite. The increase in the ratio of NAD$^+$ versus NADH reflects a distinct metabolic phenotype consistent with previously reported alterations in the energy metabolism of prostate cells. In a relatively recent study, the importance of hyperpolarized NMR probes as tools for functional studies involving the human genome was underlined by observing human cell types differing only in the mutational status of the enzyme isocitrate dehydrogenase 1 (IDH1), using hyperpolarized [1-^{13}C] alpha-ketoglutarate as a molecular reporter. IDH1 catalyzes the decarboxylation of cytosolic isocitrate to α-ketoglutarate. Specific mutations in IDH1 result in its ability to catalyze the NAD phosphate (NADPH)-dependent reduction of α-ketoglutarate to (*R*)-2-hydroxyglutarate, an onco-metabolite [81]. As a consequence, isogenic glioblastoma cells, differing only in the status of IDH1, show differences in the conversion of the hyperpolarized α-ketoglutarate to (*R*)-2-hydroxyglutarate, as probed by changing hyperpolarized NMR signal intensities.

3.2. Following Metabolism in Living Microorganisms Using Hyperpolarized ^{1}H NMR

As mentioned before, most d-DNP-based in-cell NMR studies focus on hyperpolarizing nuclei with low gyromagnetic ratios (γ), given their relatively long spin–lattice (T_1) relaxation times, e.g., non-protonated ^{13}C or ^{15}N nuclei in small molecules exhibiting short rotational correlation times (τ_c). Nevertheless, advantages can also result from observations based on hyperpolarizing and observing high-γ nuclei, e.g., protons. For example, ^{1}H signal intensities should be, on average, approximately 16-fold higher as compared with ^{13}C, given the fourfold higher gyromagnetic ratio of protons. While this gain is moderated by a concomitant increase in spectral noise, the indisputable fact that state-of-the-art ^{1}H-observation hardware is widely available and represents the most mature across all in vivo NMR technologies, might make these observations more worthwhile. Lastly, instances may arise where ^{1}H-based detection provides a better chemical discrimination than ^{13}C-based methods. In fact, spontaneous enhancements in the ^{1}H-NMR spectra of hydrogen nuclei covalently bound to hyperpolarized ^{13}C nuclei were reported a while ago [82,83]. Heteronuclear cross-relaxation effects arising in rapidly tumbling small molecules were identified as the mechanism responsible for this spontaneous polarization transfer. Given that, in such experiments, hyperpolarization can be stored in a relatively slowly relaxing nucleus that shares its hyperpolarization with a neighboring proton, opportunities arise from using these latter signals to monitor enzymatic turnover.

Using ^{1}H NMR detection, two such processes were recently studied by Frydman and co-workers to yield successful results [84] using both solutions of purified enzymes in vitro and suspensions of intact cells. The substrate in each of these studies was hyperpolarized [^{13}C] pyruvate, and the enzymatic processes targeted were (a) the production of acetaldehyde following the addition of hyperpolarized [U-^{2}H$_3$,2-^{13}C] pyruvate either to samples containing pyruvate decarboxylase (PDC) purified from *Saccharomyces cerevisiae* or, alternatively, to cultures of *S. cerevisiae* fermenting glucose, and (b) the generation of formic acid due to the activity of pyruvate formatelyase (PFL), measured in cultures of anaerobic *E. coli* following the addition of hyperpolarized [1-^{13}C] pyruvate. In these enzymatic reactions, the formation of new covalent bonds between the hyperpolarized ^{13}C nucleus and protons in the reaction products, i.e., acetaldehyde and formate, allowed the authors to transfer hyperpolarization using either insensitive nuclei enhanced by polarization transfer (INEPT)-type pulse sequences or by spontaneous cross-relaxation. Features that favored the execution of such reversed INEPT experiments included (i) the possibility to transfer magnetization between heteronuclei separated by multiple bonds (something that cross-relaxation is very inefficient at doing); (ii) the possibility of incorporating coherence-selection pulsed field gradients (PFG) to efficiently eliminate background ^{1}H signals of endogenous metabolite molecules; and (iii) the possibility of acquiring ^{1}H NMR spectra exhibiting a signal-to-noise ratio (SNR) which is approximately an order of magnitude higher than the SNR obtained using solely spontaneous magnetization transfer by cross-relaxation. In particular, using ^{1}H-detected INEPT spectroscopy allowed the detection of the acetaldehyde produced from hyperpolarized pyruvate. The authors also observed, however, that the strong signal enhancement achieved by the reversed INEPT experiment came, in part, at the expense of depleting all of the ^{13}C polarization in a single acquisition. It was, hence, deduced that lower levels of polarization transfer can be delivered using the INEPT sequence at the expense of not being able to employ more complex procedures, which would otherwise "waste" hyperpolarization generated via the DNP mechanism. It was suggested that a possible route to preserve the bulk ^{13}C hyperpolarization while using INEPT would be to use sequences that selectively excite the carbon nuclei of the product while avoiding excitation of the reactant ^{13}C coherences.

Subsequently, Frydman and co-workers explored the same biochemical process in cultures of *S. cerevisiae* using proton detection. Although these microorganisms can be grown on pyruvate as the sole source of carbon, pyruvate does not permeate the plasma membrane during glucose fermentation. Undissociated pyruvic acid, however, rapidly crosses the plasma membrane of glucose-fermenting *S. cerevisiae*. In this context, a recent d-DNP NMR study using carbon detection demonstrated that rapid diffusion of undissociated HP [1-^{13}C] acetic acid into glucose-fermenting *S. cerevisiae* occurred, in particular, at low extracellular pH [85]. After entering the cell, pyruvic acid is rapidly decarboxylated by cytosolic PDC, a process which is pronounced during the exponential stages of cell growth. The product, acetaldehyde, is present in a low equilibrium concentration and rapidly reduced to ethanol. The accumulation of acetaldehyde in *S. cerevisiae* cultures was observed with acidification of the cytosol, which was attributed to the inhibition of the enzyme alcohol dehydrogenase combined with a shift in cytosolic pH to values closer to PDC's pH optimum of 6.0 [86]. The ^{1}H NMR results obtained by Frydman and co-workers were able to corroborate this feature via detection of a relatively weak acetaldehyde ^{1}H signal found in NMR spectra of mid-exponential yeast cultures following incubation in acetate buffer at pH 4.5 and prior exposure to hyperpolarized pyruvic acid.

In a final step, the wider applicability of spontaneous polarization transfer for investigating cell metabolism was demonstrated in a study investigating the activity of PFL in *E. coli* cells. PFL rapidly metabolizes pyruvate in anaerobic *E. coli* cultures, a process that is particularly heightened when pyruvate is the main carbon source. Consistent with results obtained from direct ^{13}C detection, Frydman and co-workers observed the rapid uptake and breakdown of pyruvate as a pronounced formate ^{1}H signal. In contrast to the relatively weak acetaldehyde signal detected in *S. cerevisiae* cultures after the addition of HP pyruvic acid, the build-up and decay of the formate signal was observed on a single, scan-by-scan basis. The authors explained this higher signal-to-noise ratio by the rapid uptake of pyruvate and the accumulation of formate as a metabolic end product, characteristic of anaerobic *E. coli* cells.

In summary, Frydman and co-workers were able to demonstrate that ^{13}C-based dissolution DNP, combined with both spontaneous and INEPT-driven polarization transfer from carbon to protons, provides a clear ^{1}H signature of the enzymatic processes studied. They eventually concluded that this feature can help decipher NMR-encoded metabolic information in spectra acquired *in cellulo*. Even though metabolic fluxes from pyruvate to acetaldehyde in *S. cerevisiae* [85] and of pyruvate to formate in *E. coli* [79] were measured in the past using d-DNP-enhanced ^{13}C NMR in vitro, close signal proximity and subsequent signal overlap between the metabolic products and their direct precursors complicated these measurements in vivo to a significant extent. Also, the spontaneous transfer to ^{1}H turned out to be particularly superior to other methods, as it allowed the detection of hyperpolarized ^{1}H signals without the need for chemical manipulation of the probe molecule prior to the experiment, or modifications of the dissolution method.

4. *para*-Hydrogen-Induced Hyperpolarization Side-Arm Hydrogenation (PHIP-SAH) Method for the Detection of Cell Metabolism

As mentioned before, *para*-hydrogen-induced polarization (PHIP) is a chemistry-based hyperpolarization technique which, to a certain extent, is easier to handle and more straightforward to use when compared with DNP. This is due to the fact that (i) no additional NMR set-up extension, i.e., polarizer equipment, is needed, and (ii) polarization times are significantly shorter. The main drawback of the method in the context of in-cell NMR studies, however, is the limited availability of unsaturated precursor molecules with respect to the desired target compounds to be studied; for instance, nuclear spin-polarized acetate and pyruvate cannot be obtained by direct incorporation of the *para*-hydrogen molecule. The advent of non-hydrogenative PHIP (NH-PHIP; see below) only partially resolved this problem, given that the obtained levels of hyperpolarization in the target compounds are significantly lower as compared to using, for example, the adiabatic longitudinal transport after dissociation engenders net alignment (ALTADENA)-PHIP method. Furthermore, not all substrates of interest are suitable for NH-PHIP, given that certain requirements with respect to their molecular and electronic structure must also be fulfilled in this case.

Molecular dihydrogen occurs in the form of two nuclear spin isomers, i.e., the *ortho*- and the *para*-isomer, featuring either symmetric (triplet) or antisymmetric (singlet) nuclear spin states, respectively. Due to the three-fold degeneracy of the triplet state, *ortho*- and *para*-isomers are populated in a ratio of 3:1 under ambient conditions [87]. Interconversion between the two is symmetry forbidden and occurs, hence, at a negligibly small rate. However, when H_2 gas is flowed through an appropriate apparatus comprising a paramagnetic catalyst, i.e., activated charcoal, at low temperatures, a high enrichment of the *para* spin isomer can be achieved and observed at standard temperatures [88] due to this form being lower in energy.

When applying the *para*-hydrogen-induced polarization (PHIP) method, nuclear spin hyperpolarization is accomplished by transferring the high spin order of the *para*-hydrogen molecule to the substrate of interest. Typically, PHIP applications involve the direct incorporation of p-H_2 into unsaturated organic molecules, i.e., molecules containing either carbon–carbon double or triple bonds, in the presence of an appropriate hydrogenation catalyst. The new chemical environment experienced by the two protons upon hydrogenation breaks the singlet symmetry and renders the spin system detectable by NMR. Depending on whether the reaction is carried out at high or low magnetic field, the methods are named *para*-hydrogen and synthesis allow dramatically enhanced nuclear alignment (PASADENA) or ALTADENA, respectively [89,90]. In PASADENA experiments, the *para*-hydrogen symmetry is broken upon hydrogenation due to the distinct chemical shift environment of the two incorporated protons at high magnetic field. Thus, the NMR spectrum shows two antiphase multiplets. In ALTADENA experiments, the singlet state becomes selectively polarized, given that the chemical shifts are essentially the same when *para*-hydrogen is incorporated into the substrate at low magnetic field. In this case, the NMR spectrum is characterized by two hyperpolarized in-phase resonances of opposite sign. In the case that other magnetically active nuclei are present in the substrate, i.e., ^{13}C, ^{15}N, or ^{19}F, scalar and/or dipolar coupling interactions can cause the transfer of the initially induced proton hyperpolarization to other regions of the substrate molecule, in particular when the hydrogenation step is carried out at low field (ALTADENA); this phenomenon was also observed using other hyperpolarization methods [91,92].

Despite the relatively high degree of both homo- and heteronuclear polarization that can be achieved with PHIP, the method is not generally applicable to all molecules, i.e., metabolites, unless characteristic hydrogenation precursor modifications are applied (see below). More recently, a related methodology known as signal amplification by reversible exchange (SABRE) or NH-PHIP [93] was developed to polarize substrates without having to perform the final hydrogenation step, i.e., the physical transfer of the two protons of the dihydrogen molecule to the substrate. During a SABRE experiment, substrate and *para*-hydrogen nuclei experience transient contact interactions via the metal center of a labile catalyst–dihydrogen complex, e.g., $[Ir(H)_2(PCy_3)(substrate)_3][BF_4]$, which is formed via the reaction of $[Ir(COD)(PCy_3)(MeCN)][BF_4]$—where "Cy" is cyclohexyl and "COD" is cyclooctadiene) with *para*-H_2 and an excess of the substrate to be polarized. While the complex is formed, nuclear polarization is transferred from the *para*-hydrogen-derived protons to the molecule of interest. After the polarization transfer step, the chemically unmodified polarized substrate is released. So far, pyridine is the most widely studied SABRE substrate, and more than 10% of proton polarization was achieved in this case [94]. The method is strikingly similar to the ALTADENA experiment described above in that the polarization of the substrate occurs at low magnetic field. The magnetic field dependence of SABRE-derived signal enhancements appears to change very little with substrate type or position of the protons, suggesting that the extent of polarization depends primarily on the scalar coupling between two *para*-hydrogen-derived protons and not on the scalar coupling between *para*-hydrogen-derived and substrate protons within the hydrogenation complex [95,96].

Recently, Aime and co-workers presented a new addition to the *para*-hydrogen method, which allows the generation of high levels of nuclear polarization using substrates normally inaccessible to the *para*-hydrogen method, which can subsequently be used for hyperpolarization-assisted NMR studies in cellulo [74]. In particular, their results demonstrate that PHIP can be induced in nuclei of molecules such as acetate and pyruvate—and, in principle, other carboxylic acids as well—using precursors containing an unsaturated side-arm moiety capable of hydrogenation that can be hydrolyzed to yield the hyperpolarized target products. The reported method, named PHIP-SAH, relies on the following steps: (i) functionalization of the target acidic molecule with an unsaturated alcoholic group (i.e., vinyl or propargyl alcohol; using propargyl alcohol as a removable synthon to generate PHIP on ^{13}C resonances markedly widens the applicability of this approach); (ii) *para*-hydrogenation of the unsaturated ester; (iii) heteronuclear polarization transfer from the former pair of *para*-hydrogen protons to the carboxylate ^{13}C signal by applying magnetic field cycling, where the magnetic field is cycled between the Earth's magnetic field (hydrogenation step) and nearly zero-field (polarization transfer step) using concentric cylinders made of μ-metal, thereby increasing the extent of polarization transfer to carbon nuclei; (iv) release of the alcohol moiety via hydrolysis to obtain the polarized ^{13}C-carboxylate-containing product.

In summary, these findings open up a very interesting perspective for the use of *para*-hydrogen-based procedures for the generation of hyperpolarized, biologically relevant molecules, as demonstrated by recent NMR studies carried out in living cells using PHIP-SAH. Examples include the real-time detection of the response of the heart to altered metabolism [73], as well as the study of tumor cell metabolism using hyperpolarized ^{13}C-labeled pyruvate [72], in addition to the observation of the metabolic transformations of hyperpolarized lactate in vitro [72]. The fact that means to further enhance the efficiency of hyperpolarizing small organic molecules via PHIP-SAH were identified recently (Figure 3) will most likely lead to an even wider applicability of this very promising hyperpolarization method [97].

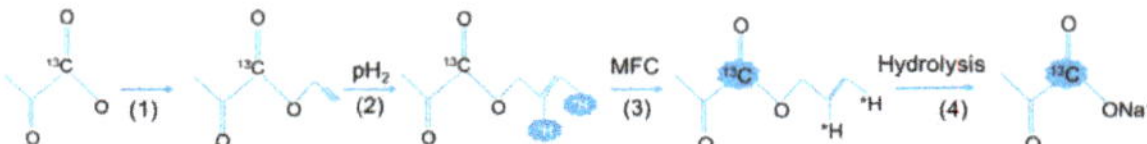

Figure 3. Diagram of the *para*-hydrogen-induced hyperpolarization side-arm hydrogenation (PHIP-SAH) procedure. (1) Functionalization of the carboxylate group with the side-arm; (2) *para*-hydrogenation of the unsaturated alcohol; (3) transfer of *para*-hydrogen spin order to the ^{13}C spin of the carboxylate group; (4) cleavage of the side-arm. The yellow background indicates reaction steps taking place in the organic phase, while the blue background indicates that the molecule is dissolved in the aqueous phase. This figure was adopted according to Reference [98].

5. In-Cell NMR and Posttranslational Phosphorylation

The most common posttranslational modifications of proteins by their respective kinases is the phosphorylation of serine, threonine, and tyrosine, as well as histidine and aspartate residues, in bacteria and fungi [99]. The addition and removal of the phosphate group by phosphatases plays a key role in the regulation of biological processes. Phosphorylation patterns modify the structure and stability of proteins, in addition to their localization and specific interactions with binding partners. Anomalous phosphorylation events are the basis for many human diseases. Thus, deciphering the phospho-code remains a major research effort.

Mass spectrometry (MS) is one of the techniques of choice to identify posttranslational modifications of the primary protein sequence due to its extreme sensitivity and high resolution. MS relies on enzymatic digestion of the protein of interest (usually with trypsin), followed by subsequent peptide analysis. However, multiple phosphorylation events which are in close proximity are difficult to analyze in terms of their exact location. MS/MS methods can solve this problem. The labile nature of the phosphate groups, however, might further hamper the analysis. Thus, mass spectrometry analyses benefit from further biochemical verifications. NMR spectroscopy is the method of choice to correlate the phosphorylation patterns with conformational changes of the protein, and in-cell NMR extends these investigations to the natural environment of the protein of interest.

5.1. In-Cell NMR within Xenopus laevis Oocytes

Selenko and co-workers utilized time-resolved NMR to monitor the sequential phosphorylation of Ser111 and Ser112 of the HIV SV40 large T antigen regulatory region [26], catalyzed by casein kinase 2 (CK2) [100]. The nuclear-import properties are modulated by these events, primarily mediated by the proximal monopartite nuclear localization sequence. The complete regulatory sequence was incorporated into the model CK2 substrate (XT 111–132 GB1). Here, the GB1 domain (B1 domain of streptococcal protein G) acts as a solubility enhancer. In the in vitro kinase assay monitored by NMR spectroscopy, CK2 firstly phosphorylated Ser112 to the maximum extent, followed by the initiation of Ser111 phosphorylation. In the S112A variant, the kinase reaction rate for the phosphorylation of S111 by CK2 was greatly reduced. Thus, phosphorylation of Ser111 depends on the pre-modification of Ser112 and the presence of a negative charge at this position. Conversely, substitution of Ser112 with aspartate revealed rapid phosphorylation of Ser111, whereas alanine at position 111 had no effect on Ser112 phosphorylation. The authors observed the same sequential phosphorylation when XT 111–132 GB1 was mixed with *Xenopus laevis* egg extract and monitored by NMR spectroscopy [20]. Finally, ^{15}N-labeled XT111–132 GB1 was microinjected into freshly prepared *Xenopus laevis* oocytes, then subsequent time-resolved in-cell NMR experiments confirmed the sequential phosphorylation of Ser112 followed by Ser111 in cellulo [26].

A second time-resolved in-cell NMR approach within *X. laevis* oocytes was reported following multiple phosphorylation of the "unique domain" of non-receptor tyrosine kinase c-Src [101]. An ^{15}N-labeled fragment of 85 residues containing the intrinsically disordered part of human c-Src was injected into the oocytes yielding an almost identical NMR spectrum compared with the in vitro spectrum with one exception: Ser17 was phosphorylated in both intact oocyte extracts. The authors then performed the time-resolved NMR experiment with extracts obtained from unfertilized *X. laevis* eggs. In this experiment, various peaks appeared/disappeared at different time points, revealing the sequential phosphorylation of three Ser residues (S17, S75, and S69). The cross-talk between the respective kinases and phosphatases was further evaluated using specific inhibitors in the oocyte extracts [101].

5.2. NMR Studies with Cell Extracts

Another NMR approach to get as close as possible to the cellular environment is to study isotope-labeled proteins in cell extracts, for example, from *Xenopus laevis* oocytes [101,102], HeLa [24,103], U2OS [24], HEK-293 [24], *Drosophila melanogaster* embryos [24], parathyroid glands [104], A2780 [103], RCSN-3 [103], B65 [103], or SK-N-SH [103]. A careful preparation preserves the proteome of the respective cells, which might even include the differentiation or cell-cycle state.

Using time-resolved NMR spectroscopy in vitro and with cell lysate, multisite phosphorylation on E26 transformation-specific (ETS)-like gene 1 (Elk-1) and corresponding extracellular signal-regulated kinase (ERK) activation was investigated [102,105]. In general, multisite phosphorylation can regulate various transcription factors involved, for example, in nuclear import and export, protein turnover, gene activation, and various protein–protein interaction (see references in Reference [105]). However, the functional role of individual phosphorylation events and their dynamics remains poorly understood. Elk-1, SAP-1, and Net form a ternary complex factor (TCF) subfamily of ETS-domain transcription factors. The kinetics of multisite phosphorylation of the Elk-1 transactivation domain (TAD) was recorded with NMR spectroscopy using recombinant ERK2 [102,106]. These time-resolved experiments revealed that phosphorylation of Thr369 and Ser384 occurred faster than the modification of Thr354, Thr364, and Ser390, whereas slower modifications were observed for residues Thr418, Ser423, and Thr337. Based upon the Michaelis–Menten constants, the authors proposed a model where an increase in the rate of phosphorylation of the "slow site" occurs after removal of fast and intermediate phosphorylation sites, which they experimentally confirmed by site-directed mutagenesis. This allowed the authors to refute certain hypotheses, for example, that fast-site phosphorylation primes later modification events [102].

Using the combination of NMR spectroscopy and cell biology, we delineated the periodic oscillation of p19^{INK4d} through the human cell cycle [24]; p19^{INK4d} undergoes a two-step phosphorylation, strictly coupled to the gap 1/synthesis (G1/S) cell phase transition. In mammalian cells, this transition is primarily regulated by a transcription factor of the E2 factor (E2F) family (E2F1 to E2F8). Most E2 factors form an inactive dimer complex with the distantly related dimerization partner (DP) protein. These E2F–DP complexes stall the cells in the G1 phase. Upon entry into S phase, the cyclin-dependent kinase complex CDK4/6–cyclin D gets hyper-phosphorylated and disrupts E2F–DP complexes. The thereby activated E2Fs are necessary for the gene expression required for the G1/S transition [107]. The kinase activity of CDK4/6 itself is inhibited by members of the CIP/KIP and INK4 protein families. The here reviewed p19^{INK4d} belongs to the latter class of inhibitors also comprising p16^{INK4a}, p15^{INK4b}, and p18^{INK4c} [108,109]. P19^{INK4d} consists of five ankyrin repeats (AR), and AR1 and AR2 bind to CDK4/6 to facilitate inhibition. Interestingly, the regulatory sites for phosphorylation (Ser66/Ser76) and ubiquitination (Lys62) are located opposite to the CDK binding interface. In order to follow phosphorylation, ^{15}N-labeled p19^{INK4d} was incubated with lysates prepared from the exponentially growing HeLa, U2OS, HEK-293 cells or from *Drosophila melanogaster* embryos. Two-dimensional (2D) ^{1}H–^{15}N HSQC-based NMR analyses indicated the pronounced chemical-shift change of the Ser66 resonance upon phosphorylation (Figure 4A). Most of the NMR resonances remained at their native positions, indicating that p19^{INK4d} largely retained its folded conformation. Ser76 did not get phosphorylated by these lysates.

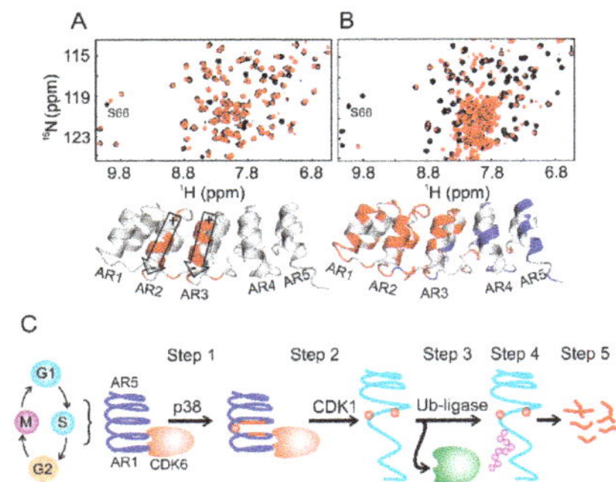

Figure 4. Following cell lysate-induced protein phosphorylation using NMR spectroscopy. (**A**) Superposition of two-dimensional (2D) ^{1}H–^{15}N HSQC spectra of phosphorylated p19^{INK4d} at Ser66 (red) and non-phosphorylated protein (black). (**B**) Superposition of 2D ^{1}H–^{15}N HSQC spectra of doubly phosphorylated p19^{INK4d} at Ser66 and Ser76 (red) and the non-phosphorylated form (black). The bottom panels in (**A**) and (**B**) represent the backbone NMR chemical-shift mapping on the p19^{INK4d} structure. (**C**) Summary of the fate of p19^{INK4d} during the cell cycle controlled by phosphorylation and ubiquitination. This figure was adapted from Reference [24].

The periodic oscillation of p19^{INK4d} during the cell cycle has its maximum during the S phase. Therefore, ^{15}N-labeled p19^{INK4d} was also incubated with S-phase-synchronized HeLa cell lysate. Surprisingly, the NMR cross-peaks, from residues belonging to AR1, AR2, and AR3, disappeared from their native position and clustered around 8 ppm along the proton dimension (Figure 4B), resembling an unfolded protein conformation. AR4 and AR5 remained folded. These structural rearrangements were induced by phosphorylation of Ser76, which turned out to be a sequential process after Ser66 phosphorylation and strictly cell-cycle-dependent. Adding specific inhibitors to the lysates before the NMR analyses identified p38 and CDK1 as kinases for Ser66 and Ser76, respectively. Further investigations, including cell-lysate NMR, revealed that phosphorylation of Ser66/Ser76 caused dissociation of the p19[INK4d]/CDK6 complex, as well as exposure of Lys62 for subsequent ubiquitination and degradation, as proposed earlier from cell biology studies [110,111]. The irreversibility of the latter step (Figure 4C) ensures, by proxy, the directionality of the cell cycle [24].

Using similar cell lysis NMR approaches [104], we disentangled the 1984 paradigm of parathyroid hormone (PTH) phosphorylation. PTH is an 84-residue peptide which controls blood calcium homeostasis. In 1984, phosphorylation of PTH was proposed based upon its high performance liquid chromatography (HPLC) elution profile being different to that of native PTH [112]. Later on, various modified forms of PTH were identified within human blood. The concentration of the modified form may rise up to 65% in patients suffering from cancer of the parathyroid glands [113–116]. In order to remain close to biological conditions, we utilized the cell lysate from bovine parathyroid gland and HEK-293 cells to incubate ^{15}N PTH. The convergent analysis of subsequent one-dimensional (1D) ^{31}P and 2D ^{1}H–^{15}N NMR, autoradiography, and matrix assisted laser desorption ionization-time of flight (MALDI-TOF) data revealed that, out of the seven serine residues of PTH, Ser1, Ser2, and Ser 17 got selectively phosphorylated. This posttranslational modification prevented in cellulo activation of PTH receptors 1 and 2 [104], which are G-protein-coupled receptors and drug targets for osteoporosis treatment [115]. PTH consists of three distinct functional regions: residues 1–17 are responsible for activation of the PTH receptors, residues ~18–34 facilitate binding to the extracellular domain (ECD) of the class B G protein coupled receptor [117–119], and the role of the intrinsically disordered residues ~34–84 might relate to the formation of functional PTH amyloids as a storage form of the hormone in secretory granules, which is currently under debate [120,121]. Our finding that phosphorylated PTH can still bind to the ECD of the PTH receptor without activation is in line with the recently reported crystal structure of the PTH1 receptor in complex with a PTH variant [115], where the N-terminus of PTH binds deeply into the trans-membrane domain of the receptor, probably resulting in the distortion of the domain by the three negative charges of phosphorylated serines 1,2, and 17.

6. Divide and Conquer by In-Cell and Cell-Lysate NMR

With an increasing complexity of proteins and protein assemblies, high-resolution structural biology methods are getting closer to their limits. One way around this is the "divide and conquer" approach [122,123], which is often applied to in vitro studies. For this, e.g., the structure, interaction sites, and/or posttranslational modifications are first determined at high resolution for isolated peptides or domains of a multi-domain protein before interpreting sparse data of the entire system. The latter might be confirmed finally by mass spectroscopy approaches [124]. "Divide and conquer" was also successful when implementing in-cell NMR and cell-lysate NMR. As described in the previous sections, isolated regulatory sites (signaling peptide sequences or isolated domains) were studied in terms of phosphorylation under in-cell/solution NMR conditions. Some examples include histone 3 modification [125], Elk-1 transactivation domain (TAD) [102], HIV SV40 large T antigen [20], N-terminal extensions to the globular protein G B1 domain [126], and N-terminal transactivation domain of human p53 [106].

While studying the conformational changes of p19[INK4d] (see previous section), the following question arose: why is the phosphorylation of Ser66 required first before Ser76 can be modified by a second kinase? When modifying a short peptide containing both serine residues by cell lysates, this sequential process was not observed (Figure 5). It turned out that Ser76 is well buried in AR3 of p19[INK4d] and that the corresponding kinase CDK1 typically phosphorylates regulatory sites in exposed loop structures [24,127]. The role of the phosphate group at Ser66 is to destabilize p19[INK4d] by repulsive interaction with the negative charge of the net dipole moments of helices 4 and 6 of AR2 and AR3. This could be shown by NMR-detected hydrogen–deuterium exchange of the backbone amides [24,128]. This thermodynamic destabilization allowed CDK1 to reach and modify Ser76.

These examples substantiate that "divide and conquer" is not only a feasible approach for in vitro structural biology, but also for in-cell NMR applications.

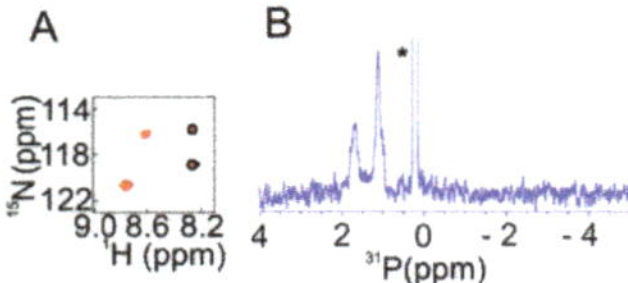

Figure 5. NMR signature of serine phosphorylation using a short peptide (A57-F86 selectively labeled with ^{15}N-Ser66 and ^{15}N-Ser76) of p19^{INK4d} modified by HeLa cell lysate. (**A**) Overlaid 2D ^{1}H–^{15}N HSQC spectra of untreated (black) and treated (red) peptide showing the typical low-field shift of the serine amide proton upon side-chain phosphorylation. (**B**) ^{31}P NMR spectra of the phosphorylated A57-F86 peptide. The asterisk in (**B**) shows the signal of the phosphate buffer.

7. Outlook of In-Cell NMR

In this review, the advances of traditional NMR virtues (structure and dynamics of proteins and hyperpolarization of small organic compounds) toward in-cell applications deciphering protein–small molecule interactions, metabolic pathways, and protein phosphorylation were summarized. The field moved from proof-of-principle to established protocols for a broad range of applications, to gain insight into the physiological and pathological processes of living cells. Despite the huge number of crystal structures in the protein databank and the "resolution revolution" in cryo-electron microscopy, one has to keep in mind that lattice-embedded proteins or vitrified biomacromolecules at very low temperatures may not reveal all of the molecule's functional properties. In-cell NMR can bridge, to some extent, these discrepancies, and further breakthroughs in the field are to be expected, especially through interdisciplinary studies, to derive general principles to tackle the functions of cellular systems and new avenues of biological research. The remaining challenges involve those regarding low concentrations and resolution. Many improving approaches were summarized using the hyperpolarization of small cellular molecules and isotope-labeled macromolecules. The combination of both improvements allowed the hyperpolarization of proteins in cell lysates [129] and of membrane proteins in living cells using solid-state NMR spectroscopy [130].

Author Contributions: A.K., L.T.K., and J.B. wrote this review.

Funding: This research was funded by the Bundesministeriums für Bildung und Forschung (ZIK HALOmem), and the Deutsche Forschungsgemeinschaft (SFB TRR 1026, BA 1821/6-1, RTG 2155).

Conflicts of Interest: The authors declare no conflict of interest.

References

1. Luchinat, E.; Banci, L. A Unique Tool for Cellular Structural Biology: In-cell NMR. *J. Biol. Chem.* **2016**, *291*, 3776–3784. [CrossRef]
2. Luchinat, E.; Banci, L. In-cell NMR: A topical review. *IUCrJ* **2017**, *4 Pt 2*, 108–118. [CrossRef]
3. Beck, M.; Baumeister, W. Cryo-Electron Tomography: Can it Reveal the Molecular Sociology of Cells in Atomic Detail? *Trends Cell Biol.* **2016**, *26*, 825–837. [CrossRef]
4. Kumar, A.; Balbach, J. Targeting the molecular chaperone SlyD to inhibit bacterial growth with a small molecule. *Sci. Rep.* **2017**, *7*, 42141. [CrossRef]
5. Luchinat, E.; Banci, L. In-Cell NMR in Human Cells: Direct Protein Expression Allows Structural Studies of Protein Folding and Maturation. *Acc. Chem. Res.* **2018**, *51*, 1550–1557. [CrossRef]
6. Serber, Z.; Keatinge-Clay, A.T.; Ledwidge, R.; Kelly, A.E.; Miller, S.M.; Dotsch, V. High-resolution macromolecular NMR spectroscopy inside living cells. *J. Am. Chem. Soc.* **2001**, *123*, 2446–2447. [CrossRef]
7. Serber, Z.; Ledwidge, R.; Miller, S.M.; Dotsch, V. Evaluation of parameters critical to observing proteins inside living Escherichia coli by in-cell NMR spectroscopy. *J. Am. Chem. Soc.* **2001**, *123*, 8895–8901. [CrossRef]

8. Serber, Z.; Straub, W.; Corsini, L.; Nomura, A.M.; Shimba, N.; Craik, C.S.; Ortiz de Montellano, P.; Dotsch, V. Methyl groups as probes for proteins and complexes in in-cell NMR experiments. *J. Am. Chem. Soc.* **2004**, *126*, 7119–7125. [CrossRef]

9. Li, C.; Wang, G.F.; Wang, Y.; Creager-Allen, R.; Lutz, E.A.; Scronce, H.; Slade, K.M.; Ruf, R.A.; Mehl, R.A.; Pielak, G.J. Protein (19)F NMR in Escherichia coli. *J. Am. Chem. Soc.* **2010**, *132*, 321–327. [CrossRef]

10. Ye, Y.; Liu, X.; Zhang, Z.; Wu, Q.; Jiang, B.; Jiang, L.; Zhang, X.; Liu, M.; Pielak, G.J.; Li, C. (19) F NMR spectroscopy as a probe of cytoplasmic viscosity and weak protein interactions in living cells. *Chemistry* **2013**, *19*, 12705–12710. [CrossRef]

11. Burz, D.S.; Dutta, K.; Cowburn, D.; Shekhtman, A. In-cell NMR for protein-protein interactions (STINT-NMR). *Nat. Protoc.* **2006**, *1*, 146–152. [CrossRef] [PubMed]

12. Burz, D.S.; Dutta, K.; Cowburn, D.; Shekhtman, A. Mapping structural interactions using in-cell NMR spectroscopy (STINT-NMR). *Nat. Methods* **2006**, *3*, 91–93. [CrossRef]

13. Maldonado, A.Y.; Burz, D.S.; Reverdatto, S.; Shekhtman, A. Fate of pup inside the Mycobacterium proteasome studied by in-cell NMR. *PLoS ONE* **2013**, *8*, e74576. [CrossRef] [PubMed]

14. Burz, D.S.; DeMott, C.M.; Aldousary, A.; Dansereau, S.; Shekhtman, A. Quantitative Determination of Interacting Protein Surfaces in Prokaryotes and Eukaryotes by Using In-Cell NMR Spectroscopy. *Methods Mol. Biol.* **2018**, *1688*, 423–444. [PubMed]

15. Xie, J.J.; Thapa, R.; Reverdatto, S.; Burz, D.S.; Shekhtman, A. Screening of Small Molecule Interactor Library by Using In-Cell NMR Spectroscopy (SMILI-NMR). *J. Med. Chem.* **2009**, *52*, 3516–3522. [CrossRef] [PubMed]

16. Bertrand, K.; Reverdatto, S.; Burz, D.S.; Zitomer, R.; Shekhtman, A. Structure of proteins in eukaryotic compartments. *J. Am. Chem. Soc.* **2012**, *134*, 12798–12806. [CrossRef] [PubMed]

17. Kubo, S.; Nishida, N.; Udagawa, Y.; Takarada, O.; Ogino, S.; Shimada, I. A gel-encapsulated bioreactor system for NMR studies of protein-protein interactions in living mammalian cells. *Angew. Chem. Int. Ed. Engl.* **2013**, *52*, 1208–1211. [CrossRef]

18. Sharaf, N.G.; Barnes, C.O.; Charlton, L.M.; Young, G.B.; Pielak, G.J. A bioreactor for in-cell protein NMR. *J. Magn. Reson.* **2010**, *202*, 140–146. [CrossRef]

19. Sakai, T.; Tochio, H.; Tenno, T.; Ito, Y.; Kokubo, T.; Hiroaki, H.; Shirakawa, M. In-cell NMR spectroscopy of proteins inside Xenopus laevis oocytes. *J. Biomol. NMR* **2006**, *36*, 179–188. [CrossRef]

20. Selenko, P.; Serber, Z.; Gade, B.; Ruderman, J.; Wagner, G. Quantitative NMR analysis of the protein G B1 domain in Xenopus laevis egg extracts and intact oocytes. *Proc. Natl. Acad. Sci. USA* **2006**, *103*, 11904–11909. [CrossRef]

21. Serber, Z.; Selenko, P.; Hansel, R.; Reckel, S.; Lohr, F.; Ferrell, J.E.; Wagner, G.; Dotsch, V. Investigating macromolecules inside cultured and injected cells by in-cell NMR spectroscopy. *Nat. Protoc.* **2006**, *1*, 2701–2709. [CrossRef]

22. Smith, M.J.; Marshall, C.B.; Theillet, F.X.; Binolfi, A.; Selenko, P.; Ikura, M. Real-time NMR monitoring of biological activities in complex physiological environments. *Curr. Opin. Struct. Biol.* **2015**, *32*, 39–47. [CrossRef] [PubMed]

23. Theillet, F.X.; Smet-Nocca, C.; Liokatis, S.; Thongwichian, R.; Kosten, J.; Yoon, M.K.; Kriwacki, R.W.; Landrieu, I.; Lippens, G.; Selenko, P. Cell signaling, post-translational protein modifications and NMR spectroscopy. *J. Biomol. NMR* **2012**, *54*, 217–236. [CrossRef] [PubMed]

24. Kumar, A.; Gopalswamy, M.; Wolf, A.; Brockwell, D.J.; Hatzfeld, M.; Balbach, J. Phosphorylation-induced unfolding regulates p19(INK4d) during the human cell cycle. *Proc. Natl. Acad. Sci. USA* **2018**, *115*, 3344–3349. [CrossRef] [PubMed]

25. Cordier, F.; Chaffotte, A.; Terrien, E.; Prehaud, C.; Theillet, F.X.; Delepierre, M.; Lafon, M.; Buc, H.; Wolff, N. Ordered Phosphorylation Events in Two Independent Cascades of the PTEN C-tail Revealed by NMR. *J. Am. Chem. Soc.* **2012**, *134*, 20533–20543. [CrossRef]

26. Selenko, P.; Frueh, D.P.; Elsaesser, S.J.; Haas, W.; Gygi, S.P.; Wagner, G. In situ observation of protein phosphorylation by high-resolution NMR spectroscopy. *Nat. Struct. Mol. Biol.* **2008**, *15*, 321–329. [CrossRef]

27. Inomata, K.; Ohno, A.; Tochio, H.; Isogai, S.; Tenno, T.; Nakase, I.; Takeuchi, T.; Futaki, S.; Ito, Y.; Hiroaki, H.; et al. High-resolution multi-dimensional NMR spectroscopy of proteins in human cells. *Nature* **2009**, *458*, 106–109. [CrossRef]

28. Banci, L.; Barbieri, L.; Bertini, I.; Cantini, F.; Luchinat, E. In-cell NMR in *E. coli* to Monitor Maturation Steps of hSOD1. *PLoS ONE* **2011**, *6*, e23561. [CrossRef]

29. Danielsson, J.; Inomata, K.; Murayama, S.; Tochio, H.; Lang, L.; Shirakawa, M.; Oliveberg, M. Pruning the ALS-associated protein SOD1 for in-cell NMR. *J. Am. Chem. Soc.* **2013**, *135*, 10266–10269. [CrossRef]

30. Ogino, S.; Kubo, S.; Umemoto, R.; Huang, S.; Nishida, N.; Shimada, I. Observation of NMR signals from proteins introduced into living mammalian cells by reversible membrane permeabilization using a pore-forming toxin, streptolysin O. *J. Am. Chem. Soc.* **2009**, *131*, 10834–10835. [CrossRef]

31. Barbieri, L.; Luchinat, E.; Banci, L. Characterization of proteins by in-cell NMR spectroscopy in cultured mammalian cells. *Nat. Protoc.* **2016**, *11*, 1101–1111. [CrossRef] [PubMed]

32. Banci, L.; Barbieri, L.; Bertini, I.; Luchinat, E.; Secci, E.; Zhao, Y.G.; Aricescu, A.R. Atomic-resolution monitoring of protein maturation in live human cells by NMR. *Nat. Chem. Biol.* **2013**, *9*, 297–299. [CrossRef] [PubMed]

33. Banci, L.; Barbieri, L.; Luchinat, E.; Secci, E. Visualization of Redox-Controlled Protein Fold in Living Cells. *Chem. Biol.* **2013**, *20*, 747–752. [CrossRef] [PubMed]

34. Hamatsu, J.; O'Donovan, D.; Tanaka, T.; Shirai, T.; Hourai, Y.; Mikawa, T.; Ikeya, T.; Mishima, M.; Boucher, W.; Smith, B.O.; et al. High-resolution heteronuclear multidimensional NMR of proteins in living insect cells using a baculovirus protein expression system. *J. Am. Chem. Soc.* **2013**, *135*, 1688–1691. [CrossRef] [PubMed]

35. Zhang, S.; Wang, C.; Lu, J.; Ma, X.; Liu, Z.; Li, D.; Liu, Z.; Liu, C. In-Cell NMR Study of Tau and MARK2 Phosphorylated Tau. *Int. J. Mol. Sci.* **2018**, *20*, 90. [CrossRef] [PubMed]

36. Primikyri, A.; Sayyad, N.; Quilici, G.; Vrettos, E.I.; Lim, K.; Chi, S.W.; Musco, G.; Gerothanassis, I.P.; Tzakos, A.G. Probing the interaction of a quercetin bioconjugate with Bcl-2 in living human cancer cells with in-cell NMR spectroscopy. *FEBS Lett.* **2018**, *592*, 3367–3379. [CrossRef] [PubMed]

37. Barbieri, L.; Luchinat, E.; Banci, L. Protein interaction patterns in different cellular environments are revealed by in-cell NMR. *Sci. Rep.* **2015**, *5*, 14456. [CrossRef]

38. Ikeya, T.; Sasaki, A.; Sakakibara, D.; Shigemitsu, Y.; Hamatsu, J.; Hanashima, T.; Mishima, M.; Yoshimasu, M.; Hayashi, N.; Mikawa, T.; et al. NMR protein structure determination in living E. coli cells using nonlinear sampling. *Nat. Protoc.* **2010**, *5*, 1051–1060. [CrossRef]

39. Luchinat, E.; Barbieri, L.; Rubino, J.T.; Kozyreva, T.; Cantini, F.; Banci, L. In-cell NMR reveals potential precursor of toxic species from SOD1 fALS mutants. *Nat. Commun.* **2014**, *5*, 5502. [CrossRef]

40. Mercatelli, E.; Barbieri, L.; Luchinat, E.; Banci, L. Direct structural evidence of protein redox regulation obtained by in-cell NMR. *Biochim. Biophys. Acta* **2016**, *1863*, 198–204. [CrossRef]

41. Ohno, A.; Inomata, K.; Tochio, H.; Shirakawa, M. In-Cell NMR Spectroscopy in Protein Chemistry and Drug Discovery. *Curr. Top. Med. Chem.* **2011**, *11*, 68–73. [CrossRef] [PubMed]

42. Prestinaci, F.; Pezzotti, P.; Pantosti, A. Antimicrobial resistance: A global multifaceted phenomenon. *Pathog. Glob. Health* **2015**, *109*, 309–318. [CrossRef] [PubMed]

43. Boucher, H.W.; Talbot, G.H.; Bradley, J.S.; Edwards, J.E.; Gilbert, D.; Rice, L.B.; Scheld, M.; Spellberg, B.; Bartlett, J. Bad bugs, no drugs: No ESKAPE! An update from the Infectious Diseases Society of America. *Clin. Infect. Dis. Off. Publ. Infect. Dis. Soc. Am.* **2009**, *48*, 1–12. [CrossRef] [PubMed]

44. Kondoh, Y.; Osada, H. High-throughput screening identifies small molecule inhibitors of molecular chaperones. *Curr. Pharm. Des.* **2013**, *19*, 473–492. [CrossRef] [PubMed]

45. Nakamoto, H.; Osada, H. Molecular chaperones as drug targets. *Curr. Pharm. Des.* **2013**, *19*, 307–308. [CrossRef] [PubMed]

46. Hadden, M.K.; Lubbers, D.J.; Blagg, B.S. Geldanamycin, radicicol, and chimeric inhibitors of the Hsp90 N-terminal ATP binding site. *Curr. Top. Med. Chem.* **2006**, *6*, 1173–11782. [CrossRef] [PubMed]

47. Löw, C.; Stubbs, M.T.; Haupt, C.; Balbach, J. Metallochaperone SlyD. In *Encyclopedia of Inorganic and Bioinorganic Chemistry*; Scott, R.A., Ed.; John Wiley & Sons, Ltd.: Hoboken, NJ, USA, 2012.

48. Scholz, C.; Eckert, B.; Hagn, F.; Schaarschmidt, P.; Balbach, J.; Schmid, F.X. SlyD proteins from different species exhibit high prolyl isomerase and chaperone activities. *Biochemistry* **2006**, *45*, 20–33. [CrossRef] [PubMed]

49. Kovermann, M.; Zierold, R.; Haupt, C.; Low, C.; Balbach, J. NMR relaxation unravels interdomain crosstalk of the two domain prolyl isomerase and chaperone SlyD. *Biochim. Biophys. Acta* **2011**, *1814*, 873–881. [CrossRef]

50. Wulfing, C.; Lombardero, J.; Pluckthun, A. An Escherichia coli protein consisting of a domain homologous to FK506-binding proteins (FKBP) and a new metal binding motif. *J. Biol. Chem.* **1994**, *269*, 2895–2901.

51. Haupt, C.; Weininger, U.; Kovermann, M.; Balbach, J. Local and coupled thermodynamic stability of the two-domain and bifunctional enzyme SlyD from Escherichia coli. *Biochemistry* **2011**, *50*, 7321–7329. [CrossRef]

52. Hiramatsu, N.; Chiang, W.C.; Kurt, T.D.; Sigurdson, C.J.; Lin, J.H. Multiple Mechanisms of Unfolded Protein Response-Induced Cell Death. *Am. J. Pathol.* **2015**, *185*, 1800–1808. [CrossRef] [PubMed]

53. Kovermann, M.; Balbach, J. Dynamic control of the prolyl isomerase function of the dual-domain SlyD protein. *Biophys. Chem.* **2013**, *171*, 16–23. [CrossRef] [PubMed]

54. Weininger, U.; Haupt, C.; Schweimer, K.; Graubner, W.; Kovermann, M.; Bruser, T.; Scholz, C.; Schaarschmidt, P.; Zoldak, G.; Schmid, F.X.; et al. NMR solution structure of SlyD from Escherichia coli: Spatial separation of prolyl isomerase and chaperone function. *J. Mol. Biol.* **2009**, *387*, 295–305. [CrossRef]

55. Kumar, A. A small-molecule acts as a 'roadblock' on DNA, hampering its fundamental processes. *J. Inorg. Biochem.* **2017**, *176*, 134–139. [CrossRef] [PubMed]

56. Harding, M.W.; Galat, A.; Uehling, D.E.; Schreiber, S.L. A Receptor for the Immunosuppressant Fk506 Is a Cis-Trans Peptidyl-Prolyl Isomerase. *Nature* **1989**, *341*, 758–760. [CrossRef] [PubMed]

57. Kawai, M.; Lane, B.C.; Hsieh, G.C.; Mollison, K.W.; Carter, G.W.; Luly, J.R. Structure-activity profiles of macrolactam immunosuppressant FK-506 analogues. *FEBS Lett.* **1993**, *316*, 107–113. [CrossRef]

58. Weissleder, R. Molecular imaging in cancer. *Science* **2006**, *312*, 1168–1171. [CrossRef]

59. Lumata, L.L.; Merritt, M.E.; Malloy, C.R.; Sherry, A.D.; van Tol, J.; Song, L.; Kovacs, Z. Dissolution DNP-NMR spectroscopy using galvinoxyl as a polarizing agent. *J. Magn. Reson.* **2013**, *227*, 14–19. [CrossRef]

60. Maly, T.; Debelouchina, G.T.; Bajaj, V.S.; Hu, K.N.; Joo, C.G.; Mak-Jurkauskas, M.L.; Sirigiri, J.R.; van der Wel, P.C.; Herzfeld, J.; Temkin, R.J.; et al. Dynamic nuclear polarization at high magnetic fields. *J. Chem. Phys.* **2008**, *128*, 052211. [CrossRef]

61. Overhauser, A.W. Polarization of nuclei in metals. *Phys. Rev.* **1953**, *92*, 411–415. [CrossRef]

62. Günther, U.L. Dynamic nuclear hyperpolarization in liquids. *Top. Curr. Chem.* **2013**, *335*, 23–69. [PubMed]

63. Miéville, P.; Jannin, S.; Bodenhausen, G. Relaxometry of insensitive nuclei: Optimizing dissolution dynamic nuclear polarization. *J. Magn. Reson.* **2011**, *210*, 137–140. [CrossRef]

64. Leggett, J.; Hunter, R.; Granwehr, J.; Panek, R.; Perez-Linde, A.J.; Horsewill, A.J.; McMaster, J.; Smith, G.; Kockenberger, W. A dedicated spectrometer for dissolution DNP NMR spectroscopy. *Phys. Chem. Chem. Phys.* **2010**, *12*, 5883–5892. [CrossRef] [PubMed]

65. Krahn, A.; Lottmann, P.; Marquardsen, T.; Tavernier, A.; Turke, M.T.; Reese, M.; Leonov, A.; Bennati, M.; Hoefer, P.; Engelke, F.; et al. Shuttle DNP spectrometer with a two-center magnet. *Phys. Chem. Chem. Phys.* **2010**, *12*, 5830–5840. [CrossRef] [PubMed]

66. Lerche, M.H.; Jensen, P.R.; Karlsson, M.; Meier, S. NMR insights into the inner workings of living cells. *Anal. Chem.* **2015**, *87*, 119–132. [CrossRef] [PubMed]

67. Schroeder, M.A.; Atherton, H.J.; Ball, D.R.; Cole, M.A.; Heather, L.C.; Griffin, J.L.; Clarke, K.; Radda, G.K.; Tyler, D.J. Real-time assessment of Krebs cycle metabolism using hyperpolarized ^{13}C magnetic resonance spectroscopy. *FASEB J.* **2009**, *23*, 2529–2538. [CrossRef] [PubMed]

68. Ball, D.R.; Rowlands, B.; Dodd, M.S.; Le Page, L.; Ball, V.; Carr, C.A.; Clarke, K.; Tyler, D.J. Hyperpolarized butyrate: A metabolic probe of short chain fatty acid metabolism in the heart. *Magn. Reson. Med.* **2014**, *71*, 1663–1669. [CrossRef] [PubMed]

69. Golman, K.; Zandt, R.I.; Lerche, M.; Pehrson, R.; Ardenkjaer-Larsen, J.H. Metabolic imaging by hyperpolarized ^{13}C magnetic resonance imaging for in vivo tumor diagnosis. *Cancer Res.* **2006**, *66*, 10855–10860. [CrossRef]

70. Bowen, S.; Hilty, C. Time-resolved dynamic nuclear polarization enhanced NMR spectroscopy. *Angew. Chem. Int. Ed. Engl.* **2008**, *47*, 5235–5237. [CrossRef]

71. Jensen, P.R.; Meier, S.; Ardenkjaer-Larsen, J.H.; Duus, J.O.; Karlsson, M.; Lerche, M.H. Detection of low-populated reaction intermediates with hyperpolarized NMR. *Chem. Commun.* **2009**, 5168–5170. [CrossRef]

72. Cavallari, E.; Carrera, C.; Aime, S.; Reineri, F. Metabolic Studies of Tumor Cells Using [1-(13) C] Pyruvate Hyperpolarized by Means of PHIP-Side Arm Hydrogenation. *Chemphyschem* **2018**. [CrossRef] [PubMed]

73. Cavallari, E.; Carrera, C.; Sorge, M.; Bonne, G.; Muchir, A.; Aime, S.; Reineri, F. The (13)C hyperpolarized pyruvate generated by ParaHydrogen detects the response of the heart to altered metabolism in real time. *Sci. Rep.* **2018**, *8*, 8366. [CrossRef] [PubMed]

74. Reineri, F.; Boi, T.; Aime, S. ParaHydrogen Induced Polarization of ^{13}C carboxylate resonance in acetate and pyruvate. *Nat. Commun.* **2015**, *6*, 5858. [CrossRef] [PubMed]

75. Meier, S.; Karlsson, M.; Jensen, P.R.; Lerche, M.H.; Duus, J.O. Metabolic pathway visualization in living yeast by DNP-NMR. *Mol. Biosyst.* **2011**, *7*, 2834–2836. [CrossRef] [PubMed]

76. Yang, C.; Harrison, C.; Jin, E.S.; Chuang, D.T.; Sherry, A.D.; Malloy, C.R.; Merritt, M.E.; DeBerardinis, R.J. Simultaneous steady-state and dynamic ^{13}C NMR can differentiate alternative routes of pyruvate metabolism in living cancer cells. *J. Biol. Chem.* **2014**, *289*, 6212–6224. [CrossRef] [PubMed]

77. Neves, A.R.; Pool, W.A.; Kok, J.; Kuipers, O.P.; Santos, H. Overview on sugar metabolism and its control in Lactococcus lactis—The input from in vivo NMR. *FEMS Microbiol. Rev.* **2005**, *29*, 531–554. [PubMed]

78. Miclet, E.; Stoven, V.; Michels, P.A.; Opperdoes, F.R.; Lallemand, J.Y.; Duffieux, F. NMR spectroscopic analysis of the first two steps of the pentose-phosphate pathway elucidates the role of 6-phosphogluconolactonase. *J. Biol. Chem.* **2001**, *276*, 34840–34846. [CrossRef]

79. Meier, S.; Jensen, P.R.; Duus, J.O. Direct observation of metabolic differences in living Escherichia coli strains K-12 and BL21. *Chembiochem* **2012**, *13*, 308–310. [CrossRef]

80. Christensen, C.E.; Karlsson, M.; Winther, J.R.; Jensen, P.R.; Lerche, M.H. Non-invasive in-cell determination of free cytosolic [NAD+]/[NADH] ratios using hyperpolarized glucose show large variations in metabolic phenotypes. *J. Biol. Chem.* **2014**, *289*, 2344–2352. [CrossRef]

81. Dang, L.; White, D.W.; Gross, S.; Bennett, B.D.; Bittinger, M.A.; Driggers, E.M.; Fantin, V.R.; Jang, H.G.; Jin, S.; Keenan, M.C.; et al. Cancer-associated IDH1 mutations produce 2-hydroxyglutarate. *Nature* **2009**, *462*, 739–744. [CrossRef]

82. Donovan, K.J.; Lupulescu, A.; Frydman, L. Heteronuclear cross-relaxation effects in the NMR spectroscopy of hyperpolarized targets. *Chemphyschem* **2014**, *15*, 436–443. [CrossRef] [PubMed]

83. Merritt, M.E.; Harrison, C.; Mander, W.; Malloy, C.R.; Sherry, A.D. Dipolar cross-relaxation modulates signal amplitudes in the (1)H NMR spectrum of hyperpolarized [(13)C]formate. *J. Magn. Reson.* **2007**, *189*, 280–285. [CrossRef] [PubMed]

84. Dzien, P.; Fages, A.; Jona, G.; Brindle, K.M.; Schwaiger, M.; Frydman, L. Following Metabolism in Living Microorganisms by Hyperpolarized (1)H NMR. *J. Am. Chem. Soc.* **2016**, *138*, 12278–12286. [CrossRef] [PubMed]

85. Jensen, P.R.; Karlsson, M.; Lerche, M.H.; Meier, S. Real-time DNP NMR observations of acetic acid uptake, intracellular acidification, and of consequences for glycolysis and alcoholic fermentation in yeast. *Chemistry* **2013**, *19*, 13288–13293. [CrossRef] [PubMed]

86. Sergienko, E.A.; Jordan, F. Catalytic acid-base groups in yeast pyruvate decarboxylase. 2. Insights into the specific roles of D28 and E477 from the rates and stereospecificity of formation of carboligase side products. *Biochemistry* **2001**, *40*, 7369–7381. [CrossRef]

87. Natterer, J.; Bargon, J. Parahydrogen induced polarization. *Prog. Nucl. Magn. Reson. Spectrosc.* **1997**, *31*, 293–315. [CrossRef]

88. Duckett, S.B.; Sleigh, C.J. Applications of the parahydrogen phenomenon: A chemical perspective. *Prog. Nucl. Magn. Reson. Spectrosc.* **1999**, *34*, 71–92. [CrossRef]

89. Bowers, C.R.; Weitekamp, D. Parahydrogen and synthesis allow dramatically enhanced nuclear alignment. *J. Am. Chem. Soc.* **1987**, *109*, 5541–5542. [CrossRef]

90. Pravica, M.G.; Weitekamp, D.P. Net NMR alignment by adiabatic transport of parahydrogen addition products to high magnetic field. *Chem. Phys. Lett.* **1988**, *145*, 255–258. [CrossRef]

91. Bargon, J.; Kuhn, L.T. Transfer of Parahydrogen-Induced Polarization to Heteronuclei. *Top. Curr. Chem.* **2007**, *276*, 25–68.

92. Kuhn, L.T. Photo-CIDNP NMR spectroscopy of amino acids and proteins. *Top. Curr. Chem.* **2013**, *338*, 229–300. [PubMed]

93. Adams, R.W.; Aguilar, J.A.; Atkinson, K.D.; Cowley, M.J.; Elliott, P.I.; Duckett, S.B.; Green, G.G.; Khazal, I.G.; Lopez-Serrano, J.; Williamson, D.C. Reversible interactions with para-hydrogen enhance NMR sensitivity by polarization transfer. *Science* **2009**, *323*, 1708–1711. [CrossRef] [PubMed]

94. Cowley, M.J.; Adams, R.W.; Atkinson, K.D.; Cockett, M.C.; Duckett, S.B.; Green, G.G.; Lohman, J.A.; Kerssebaum, R.; Kilgour, D.; Mewis, R.E. Iridium N-heterocyclic carbene complexes as efficient catalysts for magnetization transfer from para-hydrogen. *J. Am. Chem. Soc.* **2011**, *133*, 6134–6137. [CrossRef] [PubMed]

95. Adams, R.W.; Duckett, S.B.; Green, R.A.; Williamson, D.C.; Green, G.G. A theoretical basis for spontaneous polarization transfer in non-hydrogenative parahydrogen-induced polarization. *J. Chem. Phys.* **2009**, *131*, 194505. [CrossRef] [PubMed]
96. Dücker, E.B.; Kuhn, L.T.; Münnemann, K.; Griesinger, C. Similarity of SABRE field dependence in chemically different substrates. *J. Magn. Reson.* **2012**, *214*, 159–165. [CrossRef] [PubMed]
97. Cavallari, E.; Carrera, C.; Aime, S.; Reineri, F. Studies to enhance the hyperpolarization level in PHIP-SAH-produced C13-pyruvate. *J. Magn. Reson.* **2018**, *289*, 12–17. [CrossRef] [PubMed]
98. Cavallari, E.; Carrera, C.; Aime, S.; Reineri, F. (13) C MR Hyperpolarization of Lactate by Using ParaHydrogen and Metabolic Transformation in Vitro. *Chemistry* **2017**, *23*, 1200–1204. [CrossRef]
99. Walsh, C.T.; Garneau-Tsodikova, S.; Gatto, G.J. Protein posttranslational modifications: The chemistry of proteome diversifications. *Angew. Chem. Int. Ed. Engl.* **2005**, *44*, 7342–7372. [CrossRef]
100. Pinna, L.A. Protein kinase CK2: A challenge to canons. *J. Cell Sci.* **2002**, *115 Pt 20*, 3873–3878. [CrossRef]
101. Amata, I.; Maffei, M.; Igea, A.; Gay, M.; Vilaseca, M.; Nebreda, A.R.; Pons, M. Multi-phosphorylation of the intrinsically disordered unique domain of c-Src studied by in-cell and real-time NMR spectroscopy. *Chembiochem* **2013**, *14*, 1820–1827. [CrossRef]
102. Mylona, A.; Theillet, F.X.; Foster, C.; Cheng, T.M.; Miralles, F.; Bates, P.A.; Selenko, P.; Treisman, R. Opposing effects of Elk-1 multisite phosphorylation shape its response to ERK activation. *Science* **2016**, *354*, 233–237. [CrossRef]
103. Theillet, F.X.; Binolfi, A.; Bekei, B.; Martorana, A.; Rose, H.M.; Stuiver, M.; Verzini, S.; Lorenz, D.; van Rossum, M.; Goldfarb, D.; et al. Structural disorder of monomeric alpha-synuclein persists in mammalian cells. *Nature* **2016**, *530*, 45–50. [CrossRef] [PubMed]
104. Kumar, A.; Gopalswamy, M.; Wishart, C.; Henze, M.; Eschen-Lippold, L.; Donnelly, D.; Balbach, J. N-terminal phosphorylation of parathyroid hormone (PTH) abolishes its receptor activity. *ACS Chem. Biol.* **2014**, *9*, 2465–2470. [CrossRef] [PubMed]
105. Karin, M.; Hunter, T. Transcriptional control by protein phosphorylation: Signal transmission from the cell surface to the nucleus. *Curr. Biol.* **1995**, *5*, 747–757. [CrossRef]
106. Theillet, F.X.; Rose, H.M.; Liokatis, S.; Binolfi, A.; Thongwichian, R.; Stuiver, M.; Selenko, P. Site-specific NMR mapping and time-resolved monitoring of serine and threonine phosphorylation in reconstituted kinase reactions and mammalian cell extracts. *Nat. Protoc.* **2013**, *8*, 1416–1432. [CrossRef] [PubMed]
107. Wu, C.L.; Zukerberg, L.R.; Ngwu, C.; Harlow, E.; Lees, J.A. In vivo association of E2F and DP family proteins. *Mol. Cell. Biol.* **1995**, *15*, 2536–2546. [CrossRef] [PubMed]
108. Morgan, D.O. Principles of CDK regulation. *Nature* **1995**, *374*, 131–134. [CrossRef]
109. Pines, J. Cell cycle: Reaching for a role for the Cks proteins. *Curr. Biol.* **1996**, *6*, 1399–1402. [CrossRef]
110. Löw, C.; Homeyer, N.; Weininger, U.; Sticht, H.; Balbach, J. Conformational switch upon phosphorylation: Human CDK inhibitor p19INK4d between the native and partially folded state. *ACS Chem. Biol.* **2009**, *4*, 53–63. [CrossRef]
111. Thullberg, M.; Bartek, J.; Lukas, J. Ubiquitin/proteasome-mediated degradation of p19INK4d determines its periodic expression during the cell cycle. *Oncogene* **2000**, *19*, 2870–2876. [CrossRef]
112. Rabbani, S.A.; Kremer, R.; Bennett, H.P.; Goltzman, D. Phosphorylation of parathyroid hormone by human and bovine parathyroid glands. *J. Biol. Chem.* **1984**, *259*, 2949–2955. [PubMed]
113. Bringhurst, F.R. Circulating forms of parathyroid hormone: Peeling back the onion. *Clin. Chem.* **2003**, *49*, 1973–1975. [CrossRef] [PubMed]
114. D'Amour, P.; Brossard, J.H.; Rousseau, L.; Roy, L.; Gao, P.; Cantor, T. Amino-terminal form of parathyroid hormone (PTH) with immunologic similarities to hPTH(1-84) is overproduced in primary and secondary hyperparathyroidism. *Clin. Chem.* **2003**, *49*, 2037–2044. [CrossRef]
115. Ehrenmann, J.; Schoppe, J.; Klenk, C.; Rappas, M.; Kummer, L.; Dore, A.S.; Pluckthun, A. High-resolution crystal structure of parathyroid hormone 1 receptor in complex with a peptide agonist. *Nat. Struct. Mol. Biol.* **2018**, *25*, 1086–1092. [CrossRef] [PubMed]
116. Rubin, M.R.; Silverberg, S.J.; D'Amour, P.; Brossard, J.H.; Rousseau, L.; Sliney, J., Jr.; Cantor, T.; Bilezikian, J.P. An N-terminal molecular form of parathyroid hormone (PTH) distinct from hPTH(1 84) is overproduced in parathyroid carcinoma. *Clin. Chem.* **2007**, *53*, 1470–1476. [CrossRef]

117. Gardella, T.J.; Axelrod, D.; Rubin, D.; Keutmann, H.T.; Potts, J.T., Jr.; Kronenberg, H.M.; Nussbaum, S.R. Mutational analysis of the receptor-activating region of human parathyroid hormone. *J. Biol. Chem.* **1991**, *266*, 13141–13146. [PubMed]

118. Murray, T.M.; Rao, L.G.; Divieti, P.; Bringhurst, F.R. Parathyroid hormone secretion and action: Evidence for discrete receptors for the carboxyl-terminal region and related biological actions of carboxyl- terminal ligands. *Endocr. Rev.* **2005**, *26*, 78–113. [CrossRef]

119. Nussbaum, S.R.; Rosenblatt, M.; Potts, J.T., Jr. Parathyroid hormone. renal receptor interactions. Demonstration of two receptor-binding domains. *J. Biol. Chem.* **1980**, *255*, 10183–10187.

120. Evgrafova, Z.; Voigt, B.; Baumann, M.; Stephani, M.; Binder, W.H.; Balbach, J. Probing Polymer Chain Conformation and Fibril Formation of Peptide Conjugates. *Chemphyschem* **2018**. [CrossRef]

121. Gopalswamy, M.; Kumar, A.; Adler, J.; Baumann, M.; Henze, M.; Kumar, S.T.; Fandrich, M.; Scheidt, H.A.; Huster, D.; Balbach, J. Structural characterization of amyloid fibrils from the human parathyroid hormone. *Biochim. Biophys. Acta* **2015**, *1854*, 249–257. [CrossRef]

122. Gaudet, R. Divide and conquer: High resolution structural information on TRP channel fragments. *J. Gen. Physiol.* **2009**, *133*, 231–237. [CrossRef] [PubMed]

123. Wilk, T.; Fuller, S.D. Towards the structure of the human immunodeficiency virus: Divide and conquer. *Curr. Opin. Struct. Biol.* **1999**, *9*, 231–243. [CrossRef]

124. Sinz, A. Divide and conquer: Cleavable cross-linkers to study protein conformation and protein-protein interactions. *Anal. Bioanal. Chem.* **2017**, *409*, 33–44. [CrossRef] [PubMed]

125. Stutzer, A.; Liokatis, S.; Kiesel, A.; Schwarzer, D.; Sprangers, R.; Soding, J.; Selenko, P.; Fischle, W. Modulations of DNA Contacts by Linker Histones and Post-translational Modifications Determine the Mobility and Modifiability of Nucleosomal H3 Tails. *Mol. Cell.* **2016**, *61*, 247–259. [CrossRef] [PubMed]

126. Thongwichian, R.; Kosten, J.; Benary, U.; Rose, H.M.; Stuiver, M.; Theillet, F.X.; Dose, A.; Koch, B.; Yokoyama, H.; Schwarzer, D.; et al. A Multiplexed NMR-Reporter Approach to Measure Cellular Kinase and Phosphatase Activities in Real-Time. *J. Am. Chem. Soc.* **2015**, *137*, 6468–64671. [CrossRef] [PubMed]

127. Holt, L.J.; Tuch, B.B.; Villen, J.; Johnson, A.D.; Gygi, S.P.; Morgan, D.O. Global analysis of Cdk1 substrate phosphorylation sites provides insights into evolution. *Science* **2009**, *325*, 1682–1686. [CrossRef]

128. Löw, C.; Weininger, U.; Zeeb, M.; Zhang, W.; Laue, E.D.; Schmid, F.X.; Balbach, J. Folding mechanism of an ankyrin repeat protein: Scaffold and active site formation of human CDK inhibitor p19(INK4d). *J. Mol. Biol.* **2007**, *373*, 219–231. [CrossRef] [PubMed]

129. Viennet, T.; Viegas, A.; Kuepper, A.; Arens, S.; Gelev, V.; Petrov, O.; Grossmann, T.N.; Heise, H.; Etzkorn, M. Selective Protein Hyperpolarization in Cell Lysates Using Targeted Dynamic Nuclear Polarization. *Angew. Chem. Int. Ed. Engl.* **2016**, *55*, 10746–10750. [CrossRef]

130. Rogawski, R.; McDermott, A.E. New NMR tools for protein structure and function: Spin tags for dynamic nuclear polarization solid state NMR. *Arch. Biochem. Biophys.* **2017**, *628*, 102–113. [CrossRef]

International Journal of
Molecular Sciences

MDPI

Review

Quo Vadis Biomolecular NMR Spectroscopy?

Philipp Selenko

Weizmann Institute of Science, Department of Biological Regulation, 234 Herzl Street, Rehovot 76100, Israel;
philipp.selenko@weizmann.ac.il; Tel.: +972-(0)8-934-2991

Received: 16 February 2019; Accepted: 8 March 2019; Published: 14 March 2019

Abstract: In-cell nuclear magnetic resonance (NMR) spectroscopy offers the possibility to study proteins and other biomolecules at atomic resolution directly in cells. As such, it provides compelling means to complement existing tools in cellular structural biology. Given the dominance of electron microscopy (EM)-based methods in current structure determination routines, I share my personal view about the role of biomolecular NMR spectroscopy in the aftermath of the revolution in resolution. Specifically, I focus on spin-off applications that in-cell NMR has helped to develop and how they may provide broader and more generally applicable routes for future NMR investigations. I discuss the use of 'static' and time-resolved solution NMR spectroscopy to detect post-translational protein modifications (PTMs) and to investigate structural consequences that occur in their response. I argue that available examples vindicate the need for collective and systematic efforts to determine post-translationally modified protein structures in the future. Furthermore, I explain my reasoning behind a Quinary Structure Assessment (QSA) initiative to interrogate cellular effects on protein dynamics and transient interactions present in physiological environments.

Keywords: in-cell NMR; time-resolved NMR; post-translational modifications; structure function; intrinsically disordered proteins

---

1. Preface I

In-cell nuclear magnetic resonance (NMR) spectroscopy has become increasingly popular amongst the biomolecular NMR community, especially in funding statements regarding the versatility of general NMR methods or to embellish visionary outlooks in grant applications of high-risk caliber. In practice, genuine in-cell NMR measurements remain sparse and few laboratories are committed to performing these experiments. As a consequence, primary in-cell NMR work is all too often outnumbered by recurring reviews [1–3]. To not perpetuate this trend, I refrain from discussing in-cell NMR work itself but rather focus on spin-off applications whose developments were stimulated by cellular NMR approaches. As we shall see, these alternatives offer enticing possibilities for the biologically inclined NMR spectroscopist, especially in light of the resolution revolution in cryo-electron microscopy (EM) and its impact on all areas of structural biology [4].

2. Preface II

The arrival of cryo-EM in the realm of genuine atomic-resolution methods has changed the structural biology landscape in a most profound manner. While X-ray crystallographers adapted swiftly to this new reality, largely by abandoning their methodological preferences altogether, the biomolecular NMR community has been slow in accepting what is evidently more than a fleeting trend. Without a doubt, the future of structural biology will be shaped by EM-based methods with recent breakthroughs offering a mere glimpse into the great potential of the technique. This holds true for advancements in EM technologies, as well as for turning previously intractable biological questions into feasible research projects. By agreeing that EM will dominate the field of structural biology in the next decades, we may ask what questions the community will address with it? Obvious answers include large biomolecular complexes and machines, reconstituted and native membrane proteins, macromolecular protein assemblies including oligomers, fibrils and ordered aggregates, especially those purified from natural sources, as well as low abundance specimens and samples [5]. While such targets may initially be chosen based on their structural rigidity and size, these limitations will likely shift to smaller scales and more flexible assemblies in the near future. Beyond these self-evident trends, another popular theme in general biology is also making its mark in structural biology—the goal to investigate biomolecules where they naturally occur, that is, inside cells. In most cases, this means to derive structural and functional insights directly in intact cellular specimens. Accordingly, the term cellular structural biology was coined to collectively describe such high-resolution in situ efforts [1,6–8]. X-ray crystallography is mostly excluded from cellular applications due to the need for crystalline samples, which are (almost) impossible to obtain inside cells [9]. By contrast, the history of electron microscopy is firmly rooted in cellular investigations and precedes high-resolution structural studies by decades. Concomitantly, technical developments in cell- versus molecule-based EM methods ran parallel for many years and converged only recently. This was largely due to technical advancements including direct electron detectors, fast imaging modalities and computational 'drift' corrections, improved energy filters, and phase-plate technologies for greater contrast [10,11]. These developments offered mutual benefits to both branches of EM methods by alleviating common bottlenecks. Today's breakthroughs in cellular cryo-electron tomography (CET) [12–14], cryo-scanning transmission electron tomography (CSTET) [15,16], and correlative light and electron microscopy (CLEM) [17–19] may seem synonymous with advances in single particle cryo-EM, although, strictly speaking, they developed independently. CET, CSTET, and CLEM represent the core tools of current EM methods in cellular structural biology, with most insights generated by CET of vitrified prokaryotic [20] or focused-ion beam (FIB) milled eukaryotic cells [21], commonly by placing previously determined, high-resolution structures into subtomogram-averaged densities [22]. This will likely change in the future and three-dimensional architectures of proteins and their complexes will be solved directly in cells. These outlooks serve as the preamble for all structural in situ studies today, including those by in-cell NMR spectroscopy. In comparison, past, present, and future in-cell NMR efforts are at a sizable disadvantage.

3. Pitfalls, Challenges, and Opportunities

Let me first survey the historical context of modern biomolecular NMR applications and their standing in the aftermath of the EM revolution in resolution. Throughout the 1990s, and following the discovery of independently folded protein domains as key functional units in many biological processes, NMR spectroscopy and X-ray crystallography often pursued these targets in a friendly competitive manner [23,24]. Based on the growing awareness that protein domain functions can only be appreciated in their extended modular contexts, ever more sophisticated NMR approaches, often combined with complementary techniques such as small-angle X-ray scattering (SAXS) aimed at determining structures of large multi-domain constructs [25]. Faced with the dilemma of exponential signal multiplication with target size and correspondingly adverse relaxation properties, the biomolecular NMR community devised several ingenious ways to deal with such samples. These include (transverse) relaxation-optimized spectroscopy (TROSY) [26,27], amino acid-specific, site-selective, and segmental isotope labeling [28], higher-dimensionality NMR experiments [29], and non-uniform sampling schemes [30]. Despite these developments, structure determination efforts of large, folded proteins by solution-state NMR spectroscopy remained challenging and cumbersome. Luckily, the emergence of structural disorder as a fundamental property of higher eukaryotic proteomes presented itself as the timely savior from this failing arms race [31,32]. Members of the biomolecular solution NMR community, including us, converted to the safe haven of intrinsically disordered proteins (IDPs) in large numbers and with great fervor [33–36]. In the absence of natural predators, solution NMR flourished in its newly discovered habitat and, emboldened by previous advancements that proved equally if not more useful in the world of dynamic disorder, it experienced a formidable revival [37]. Nonetheless, some conceptual problems persisted, which, I believe, were often rooted in the classically trained mindsets of its 'structured' protagonists and a certain negligence of the biological implications of disorder with regard to function rather than to structure (or absence thereof). By yet another stroke of good luck, liquid–liquid phase separation and membraneless organelles emerged as the timely saviors of IDP–NMR scientists in the midst of their structure–function identity crises [38,39]. Suddenly, the absence of structure became 'function' and a thorough discussion about the general role of disorder in biological systems could safely be postponed.

In parallel to these topical developments, solid-state (ss) NMR spectroscopy matured into a veritable contestant for structural analyses of high-molecular-weight assemblies of ordered proteins, including viral particles [40], amyloid fibrils [41], and membrane proteins [42]. Emboldened by similarly brazen developments in ssNMR methods, including proton-based excitation and detection routines [43], dynamic nuclear polarization (DNP) [44], and probes for magic angle spinning (MAS) at ultra-fast frequencies (>100 kHz) [40], biological ssNMR applications reached unprecedented levels of sophistication. However, many high-quality ssNMR samples were equally good cryo-EM specimens, which led to both methods often pursuing the same targets. This is illustrated by complementary ssNMR and cryo-EM studies on amyloid proteins such as Het-S [45–49], amyloid-β [50–53], α-synuclein [54–62] and tau [63–68], viral particles [69–71], membrane proteins including bacterial secretion systems [72–76], and transporters [77–79]. Once again, the competitive outcomes of these efforts proved unfavorable for NMR spectroscopy, especially in terms of cost effectiveness, required sample types and quantities, experimental times, and computational efforts to turn measured distance restraints into high-resolution, structural models. In addition, ssNMR often provided restricted information about parts of these multicomponent machineries, whereas cryo-EM revealed comprehensive insights into entire assemblies. Moreover, CET and cryo-EM methods were beginning to resolve these structures in their native cellular settings, as recently shown for the type III secretion system [80,81] or with specimens purified directly from post-mortem brains as in the case of human tau filaments [66–68], respectively. These breathtaking developments consolidated a similarly critical assessment of solid-state NMR spectroscopy by many structural biology departments turning to cryo-EM as their preferred option for future investments instead.

In-cell NMR spectroscopy entered this stage about two decades ago [82]. Despite the largely benevolent patronage of the biomolecular NMR community, it developed slowly even though it offered a principally attractive means to populate a structural biology niche that no other method could claim at that time: The ability to obtain atomic-resolution insights into protein, RNA, and DNA structures directly in live cells. While in-cell NMR was nourished with good intentions, most biomolecular NMR laboratories chose to observe it from a distance rather than getting involved themselves. After all, those were the times of NMR resources and projects abound and dabbling in unknown territories such as cells required skills that were too remote for many to bother with. As a consequence, the early days of in-cell NMR were spent in comfortable solitude. Over the years, and having transcended from bacteria to higher eukaryotic and mammalian cells, interest within the community grew. Encouraged by more frequent high-impact publications and dedicated in-cell NMR sessions at regular NMR meetings, additional groups began to explore the method. With times changing, cryo-EM looming, and NMR resources dwindling, the prospects of in-cell NMR, including the comfort of a protected niche, are becoming ever more appealing. Despite this positive trend, true in-cell NMR applications remain challenging and initial experiments often fail to produce the desired outcomes (also for reasons that recent in-cell NMR work has helped to elucidate, see below). To minimize such drawbacks, I advise novice users to follow three simple rules when embarking on in-cell NMR adventures. First, start with crude lysates of your 'empty' host cells and add isotope-labeled proteins/biomolecules directly to these slurries for pilot NMR experiments. Possible interactions with cellular components will be recapitulated in these mixtures and degrees of (site-selective) line broadening will be indicative of scenarios to be encountered in cells (see also the quinary structure section at the end of this text). Second, set up stringent control experiments to ensure that enrichment or delivery of isotope-labeled biomolecules indeed produces cellular samples that harbor the species of interest. Third, maximize efforts to quantify the amount of target biomolecules in these samples in order to (a) repeat lysate experiments at relevant (low µM) concentrations and (b) optimize enrichment or delivery procedures for satisfactory signal-to-noise ratios within realistic data acquisition times. Guidelines for how to measure intracellular concentrations of delivered/enriched biomolecules in cells are provided in several published protocols [83–86]. Together, these steps will allow newcomers to pre-assess the overall feasibility of in-cell NMR projects.

As I said before, the goal of this review is not to discuss in-cell NMR applications but to focus on concepts and approaches that the method helped to popularize. Similar to in-cell NMR, these are principally rooted in five fundamental aspects of NMR spectroscopy: One, NMR is an atomic-resolution method. Two, NMR is quantitative in that signals reflect the number of NMR-active nuclei in the sample. Three, NMR resonance frequencies (chemical shifts) are uniquely sensitive to the chemical environment of active nuclei, which provides structural and functional information. Four, NMR signals contain additional information about the dynamic properties of observable spin systems. Five, NMR spectroscopy is a non-destructive method and works at physiological temperature and pH. As we shall see next, this minimal set of key NMR properties provides the basis for genuine applications off the paths of conventional structural biology routines.

4. Detecting Post-Translational Protein Modifications by NMR

In cells, most proteins undergo different post-translational modifications that alter the chemical identities of individual residues [87]. The establishment of PTMs is strictly regulated and mediated by dedicated sets of enzymes, whose own activities are often controlled by PTMs. Together they form cascading feedback networks of interconnected signaling pathways. Importantly, most PTMs constitute reversible 'switches' that reprogram the functions of proteins in response to external and internal cellular cues. Therefore, they serve as key regulators in virtually all processes of life. PTMs typically occur at multiple sites and in disordered protein regions [88,89]. Their addition and removal are highly dynamic and combinatorial modification patterns are often established in multistep reaction mechanisms with clearly defined hierarchies [90]. Given the immense importance of PTMs in modulating protein functions, biologists seek to understand where PTMs occur, which types of modifications are present, and in what combinations, how they are established (and removed), and how they impact the biological activities of target proteins. The breadth of these questions explains the great need for analytical tools to annotate and investigate protein PTM states.

The unique sensitivity of the NMR chemical shift on its immediate chemical environment provides an exquisite readout modality for PTM states of individual protein residues. In other words, changes in the chemical composition of protein side-chain moieties, as imposed by different PTMs, result in chemical shift changes that are characteristic of the respective type of modification (see [91] for a comprehensive overview). In turn, differences in protein–NMR spectra reveal the sites of PTMs as well as their nature, which constitutes a highly attractive, analytical feature. In 2008, we reported how in-cell NMR can be used to detect and site-specifically assign cellular protein phosphorylation events, the most common type of eukaryotic PTM [92]. While our publication was not the first to describe the detection of protein phosphorylation by NMR spectroscopy, and was clearly inspired by earlier work by Lippens, Gronenborn, Forman-Kay, and others [93–95], it probably boosted renewed interest in using NMR to decipher more complex modification reactions and their underlying mechanisms. One aspect of our paper that may have additionally stimulated this trend was the use of cell lysates containing native cellular enzymes and modification reactions that we carried out directly in these lysates, thus offering a cost-effective and convenient way to phosphorylate isotope-labeled proteins for NMR measurements in vitro and in situ [96]. In the following years, the rationale for detecting protein phosphorylation by NMR spectroscopy was successfully employed by many groups (see below). Importantly, most of these studies were aimed at deriving functional insights rather than at delineating structural information about modified substrate states. By exploiting the non-destructive and quantitative nature of such NMR measurements, we illustrated another analytical advantage of this approach: The ability to directly follow PTM reactions in a time-resolved fashion in order to deduce site-specific modification rates [97,98]. Indeed, protein phosphorylation studies by time-resolved NMR spectroscopy became very popular and proved essential for delineating mechanistic insights into diverse sets of signaling reactions [99–135].

As stated earlier, NMR detects PTMs irrespective of their nature. Moreover, different PTM chemistries impose characteristic spectral signatures, which serve to identify the corresponding modification type(s) [91]. Therefore, functional studies of protein phosphorylation represent just one aspect of NMR's power as an analytical tool. Indeed, several other protein PTMs have been analyzed by NMR spectroscopy, including acylation, alkylation, and glycosylation [136–138]. More recently, we used time-resolved NMR spectroscopy to study methionine oxidation as an example of a non-enzyme mediated protein PTM, which occurs frequently in response to oxidative cell stress and organismal aging [139]. Specifically, we monitored the site-selective repair of the oxidation-damaged α-synuclein by endogenous cellular enzymes. Differences in NMR chemical shifts of native and oxidation-damaged proteins allowed us to follow individual repair reactions in a time-resolved manner with single-residue resolution. Similarly, we performed time-resolved in situ NMR measurements to monitor irreversible protein cleavage in isolated protein–protease mixtures, cell lysates, and intact cells [140]. Together, these examples underscore the great potential of NMR-based PTM studies in reconstituted in vitro systems ranging from defined enzyme-substrate mixtures to native, cell-free lysates. They also paved the way for applications that I will discuss next: NMR investigations of changes in protein structures upon post-translation modifications.

5. Making and Breaking of Protein Structures by PTMs

In the previous paragraph, I outlined how NMR spectroscopy can be employed to functionally annotate different types of PTMs in a residue- and time-resolved, quantitative manner. While such applications exploit the analytical power of NMR spectroscopy, they offer an attractive additional feature: The ability to correlate the establishment of different PTMs with structural alterations that occur in their response. Phosphorylation in particular has long been known for its capacity to alter protein structures [141], especially by strengthening or weakening secondary structure elements such as α-helices in a position-dependent manner [142]. Whereas phosphorylation of N-terminal helix residues stabilizes helicity via capping interactions, modifications at C-terminal helix residues add to the negative dipole moment at helix ends and act destabilizing [143]. Many examples of such stabilizing and destabilizing phosphorylation effects in classically folded proteins are known [144]. The emergence of IDPs has helped to extend this concept to regions of residual secondary structure, where phosphorylation appears to exert even greater effects [93,108,145]. Given that pre-structured motifs usually function as molecular recognition elements in IDP–ligand interactions [146], signaling and phosphorylation-dependent changes in these structural propensities influence binding energies and affinities in pronounced ways, especially in binding-induced disorder-to-order and order-to-disorder transitions [147]. Systematic structural investigations into these types of IDP–PTM interactions are scarce [148], although I believe that they provide important new insights into the roles of PTMs in signaling-mediated structure–function relationships. At this point, I wish to reiterate that solution NMR is uniquely capable of providing this information, especially with regard to dynamic and partially disordered protein regions, where most eukaryotic PTMs occur. No other atomic-resolution method can unravel these scenarios at comparable levels of resolution and with similar ease. Indeed, several publications pay tribute to the great power of NMR in such structure–function analyses [94,100,105,149–159]. In the following paragraphs, I discuss three examples that illustrate the scope of phosphorylation-induced structural rearrangements and NMR's excellent ability to decipher them, and their resulting architectures.

My first example is the human splicing factor 1 (SF1). Together with the large and small subunits of the U2 small nuclear ribonucleoprotein auxiliary factor (U2AF), SF1 defines the 3′ splice site recognition complex on pre-messenger RNA. While SF1-U2AF binding is primarily mediated by a canonical tryptophan–RNA recognition motif (RRM) interaction [160], the ternary RNA–protein complex is further stabilized by SF1 phosphorylation at two adjacent residues [161]. In 2013, two groups independently reported the molecular basis for this behavior by determining the X-ray and NMR structures of phosphorylated and unmodified SF1-U2AF complexes [162,163]. The main reason for the importance of these publications is that they provide complementary information about the structural effects of SF1 phosphorylation that are inaccessible to either experimental method alone. Hence, in this case, the combination of X-ray crystallography and NMR spectroscopy revealed the full scale of the phospho-regulation of the SF1-U2AF interaction. Residues in the N-terminus of SF1 arrange in a helix-hairpin conformation with a flexible ~30-residue linker connecting the two α-helices. In unmodified SF1, this intrinsically disordered region (IDR) exhibits high internal dynamics and samples a range of conformations [163]. In turn, the IDR of unmodified SF1 is poorly defined by X-ray crystallography with missing electron density for most of its residues [162]. By contrast, the very same residues were perfectly tractable by NMR and their dynamic properties annotated with high precision [163]. Upon phosphorylation of the two SF1 serines, the IDR becomes fixed in a rigid conformation, primarily via coordination of the two phosphate moieties by conserved arginines at the N-terminus of the second SF1 helix. Accordingly, the X-ray structure provided detailed insights into the coordination of phosphate oxygens by side-chain guanidium groups in a tight arginine 'claw' [162]. NMR spectroscopy, on the other hand, measured greatly reduced dynamics of the phosphorylated linker and the formation of a stable, albeit disordered structure [163]. Thus, the combination of both methods revealed how dual phosphorylation of two SF1 linker sites locked the IDR in a conformation that cooperatively enhanced U2AF binding via reducing the entropic penalty of the encounter complex. The SF1-U2AF example provides several important lessons: First, it outlines how the regulated establishment of intramolecular IDR contacts can drastically alter the binding behavior of two proteins. Second, it illustrates how cell signaling and protein phosphorylation regulate this behavior in a fully reversible manner. Third, it underscores the importance of novel types of intramolecular coordination chemistries between PTMs and protein residues in forming previously unknown structures such as the arginine–phosphate claw observed here. Given the abundance of protein phosphorylation throughout eukaryotic biology, such reversible PTM structures likely constitute common themes in many signaling processes. Intramolecular arginine–phosphate contacts, for example, are present in a large number of substrates and clearly qualify as a general structural principle [159,164,165]. Despite this, only a handful of PTM motifs are known to adopt defined conformations in their modified states [166], with kinase domains and their phosphorylated activation loops serving as the most prominent examples [167]. Accordingly, our understanding about sequence features giving rise to such structures is limited and we are unable to predict when and where they occur, or what types of conformations they adopt. Therefore, I believe that the conformational space of possible three-dimensional protein 'folds' is much larger than we think and that we will only grasp its full dimensions when we begin to analyze the structures of post-translationally modified proteins in a systematic manner. Given that disorder and high degrees of protein flexibility will likely prevail in these uncharted territories, NMR spectroscopy is ideally suited to uncover novel PTM structures.

Along those lines, I discuss here another extraordinary example of PTM-induced protein folding: The eukaryotic translational initiation factor 4E (eIF4E) binding protein, 4E-BP. In its non-phosphorylated form, isolated 4E-BP is fully disordered [168,169]. In the presence of eIF4E, 4E-BP adopts a helical conformation and binds eIF4E via a conserved, hydrophobic interaction motif [170]. Surprisingly, phosphorylated 4E-BP fails to interact with eIF4E [171]. In 2015, a concerted NMR effort revealed stunning insights into the mechanistic basis for this behavior. Kay and Forman-Kay, et al. showed that signaling-mediated modification of 4E-BP at two threonine residues upstream the hydrophobic interaction motif induced complete folding of the eIF4E cognate site into a binding-incompetent β-sheet structure [172]. Thus, phosphorylation of 4E-BP switches the protein between a disordered and a β-strand conformation, in which the eIF4E binding site is inaccessible. By doing so, 4E-BP defines a functional signaling mechanism that is entirely different in its mode of action. Rather than to establish a recruitment platform for phospho-binding proteins, phosphorylation masks an existing interaction motif by virtue of incorporating it into a folded structure. Therefore, the 4E-BP example of phosphorylation-induced protein folding adds a novel structural dimension to eukaryotic signaling processes. Probably the most radical aspect of this and the previously discussed SF1 case is the direct involvement of phosphates in forming the core contacts of the newly formed structures [162,172]. Both examples reveal radical new principles of how phosphate coordination gives rise to globular protein folds without classical hydrophobic cores [173]. Built from disorder, these structures also suggest new paradigms for biological regulation. Phosphorylation-induced protein folding may shield existing binding sites, or create new ones. It may expose critical residues for ubiquitination and, thereby, trigger cellular degradation, or act in the opposite direction and prevent protein turnover, thus extending the lifetimes of molecular players. Most importantly, SF1 and 4E-BP exemplify new modes of biological regulation and remind us that many of them remain to be discovered. In addition, both provide compelling testimonies to the power(s) of NMR in deciphering such complex structure–function relationships.

Finally, I outline an example in which phosphorylation triggers the opposite effect in that it unfolds a folded protein domain [174,175]. Specifically, I discuss how phosphorylation of the human cell-cycle regulator and cyclin-dependent kinase inhibitor p19^{INK4d} drives the transition from G1 to S-phase in an irreversible manner [176]. In contrast to phosphorylation-induced folding of 4E-BP, dual phosphorylation of p19^{INK4d} acts to dissolve its central ankyrin-repeat domain in a stepwise manner. While modification of the first residue destabilizes N-terminal helices to provide access for a second enzyme to phosphorylate a previously inaccessible site, establishment of both modifications unfolds the entire N-terminus of p19^{INK4d}. This exposes conserved lysine residues for ubiquitination, which, in turn, triggers p19^{INK4d} degradation by the proteasome and clearance from cells. At the same time, the two-step cascade also abolishes the structured CDK6 binding interface on p19^{INK4d}, which disrupts the inhibitory CDK6-p19^{INK4d} complex, releases CDK6, and activates it to signal G1/S transition. Thus, p19^{INK4d} phosphorylation, unfolding, and degradation collectively act to ensure the directionality of cell-cycle progression, probably one of the most important processes in eukaryotic biology. The beauty of this mechanism, and the corresponding study, lies in the exquisite combination of NMR spectroscopy and cell biology methods, which, I believe, serve as a paradigm for future investigations in these directions. NMR experiments were performed with isolated, recombinant enzymes as well as in lysates of cells arrested at different stages of the cell cycle and containing different sets of active, endogenous kinases. Structure–function analyses combined with biochemical pull-down and protein detection assays revealed the phosphorylation-triggered dissociation of the CDK6-p19^{INK4d} complex, subsequent ubiquitination, and cellular clearance of p19^{INK4d}. All in all, a benchmark study of how biomolecular NMR spectroscopy may be used in combination with complementary biochemistry techniques to reveal novel modes of biological regulation. The overall paucity of PTM-induced folding and unfolding examples may have two reasons: Either they are rare, or we have not used adequate tools to uncover them on a broader scale. I am convinced of the latter and believe that NMR spectroscopy can serve as a key discovery technique in these investigations.

6. Physiological Protein Dynamics and Quinary Structure

Finally, I want to touch upon a subject that I feel is underrepresented in NMR studies aimed at resolving physiological protein behaviors. On the one hand, NMR is uniquely capable of providing quantitative information about protein dynamics over time scales spanning several orders of magnitude, from pico-seconds (10^{-12} s) to hours (~10^4 s) [177–179]. On the other hand, most NMR relaxation measurements are performed on isolated, dilute samples that bear little resemblance to the cellular environments where proteins function [180]. As a result, we gathered a wealth of information about protein dynamics in artificial in vitro settings without understanding how they are manifested in vivo. I believe that this has led to several misconceptions about relevant time scales of protein motions in cells [37]. I agree that comprehensive NMR relaxation studies under true in-cell conditions are difficult to perform. I also acknowledge that in vitro settings that approximate cellular environments often fail to recapitulate physiological protein behaviors [181]. One reason for this shortcoming lies in the overwhelming complexity of the intracellular milieu, both in terms of composition and organization. Accordingly, in cells, proteins experience highly diverse sets of encounters and continuously engage in transient, low-affinity associations that may outnumber specific binding events in abundance and frequency. These interactions are often protein and cell type-specific and appear to have been evolutionarily optimized across entire proteomes [182,183]. Despite the randomness of these short-lived encounters, their configurations, orientations, and binding surfaces tend to deviate from stochastic behaviors. Specifically, properties such as charge distributions and local hydrophobicity steer association characteristics that are not uniform across protein surfaces [184]. As a result, some areas display higher propensities for unspecific binding events, or, more accurately, they mediate interactions of extended residence lifetimes. With respect to protein dynamics, this results in non-uniform attenuations that cannot be recapitulated with single crowding or viscosity agents. The nature of these effects in cells, both specific and unspecific, and their combined influence on protein structure and dynamics is summarized in the term "quinary protein structure"—in extension to the classical definition of primary, secondary, tertiary, and quaternary protein structures. While quinary protein structure may be considered a modern concept, it was first coined in the 1980s without much traction for the remainder of the century because tools to assess quinary structure in cells did not exist [185–187]. Modern in situ methods including in-cell Förster resonance energy transfer (FRET) and NMR spectroscopy helped to change this notion, and contributed to a formidable revival of the term, with several studies emphasizing its rediscovered biological importance [102,183,188–195]. One general conclusion from these investigations is that intracellular environments and respective quinary structure interactions can have opposite effects and may act stabilizing or destabilizing in a protein-specific manner [196]. Another observation that we and others made is that partially structured motifs and regions of residual structure, synonymous with protein interaction sites, are usually more prone to exhibit quinary structure effects than biologically inert parts of proteins, as expected for functional but uncomplemented electrostatic and hydrophobic surfaces [197]. In turn, quinary interactions at these sites attenuate NMR relaxation properties and respective signal qualities to greater extents than cellular viscosity and crowding alone. Such effects serve as useful indicators for regions of biological interest and their identification may aid the discovery of new interaction hotspots even when the types and identities of interacting ligands are not known. Given the great predictive value of this information, I propose to launch a concerted *Quinary Structure Assessment (QSA)* initiative to annotate the impact of intracellular environments on protein structures and dynamics. To do so, I invite interested NMR groups to record 2D ^{1}H-^{15}N NMR spectra of their favorite proteins in a standard buffer and in easily accessible cell lysates prepared from bacteria (*E. coli*) [198] and a mammalian cell line such as HeLa (*H. sapiens*) [98]. By acquiring reference and lysate NMR spectra with identical spectrometer settings, they will be able to extract information about site-selective line broadening. Even when resonance assignments are not available, such comparisons will provide qualitative and quantitative information about altered relaxation properties due to quinary structure interactions. Sharing these results with dedicated in-cell NMR laboratories such as ours will allow us to formulate multi-component mixtures containing metabolites, RNA, DNA, proteins, and lipids, in order to reproduce the observed cellular effects under stable in vitro conditions. Stock solutions of these mixtures may then be distributed amongst the biomolecular NMR community to collectively probe quinary structure interactions in other protein samples, which will reveal fundamental insights into core aspects of cellular biology and biophysics. As an added value, they will provide general feasibility assessments for in-cell NMR approaches with individual target proteins (see recommendations for pilot in-cell experiments above).

7. Conclusions

In summary, I have presented a collective outlook towards the implementation of biomolecular NMR methods to investigate basic biological processes under experimental conditions that

(a) approximate native cellular settings, (b) exploit endogenous machineries such as enzymes, and (c) follow their activities in a time-resolved and quantitative fashion. Specifically, I outlined the use of solution NMR spectroscopy to study post-translational protein modifications (PTMs) and their effects on protein structure and dynamics, and to perform time-resolved NMR analyses of enzyme-substrate reactions to determine modification rates, processing mechanisms, as well as PTM hierarchies and cross-talks. Furthermore, I introduced quinary structure interactions and how they affect in-cell protein dynamics. Importantly, most of the methodological aspects discussed in this review concern NMR experiments performed outside of cells, with standard protein samples, spectrometer setups, and settings. As I hope to have conveyed to the reader, I firmly believe that these NMR applications offer exciting new research directions beyond classical structure determination routines. In this day and age, we, the biomolecular NMR community, are well advised to explore these non-conventional paths with vigor, rigor, and timely urgency.

Funding: This work was funded by the ERC Consolidator Grant NeuroInCellNMR (647474) to P.S.

Acknowledgments: I thank Sharon Greier Wolf for discussions on cellular cryo-EM methods and comments on the manuscript.

Conflicts of Interest: The author declares no conflict of interest.

Abbreviations

NMR	Nuclear Magnetic Resonance
EM	Electron Microscopy
CET	Cryo-Electron Tomography
CSTET	Cryo-Scanning Transmission Electron Tomography
CLEM	Correlative Light and Electron Microscopy
TROSY	Transverse Relaxation Optimized Spectroscopy
MAS	Magic Angle Spinning
DNP	Dynamic Nuclear Polarization

References

1. Luchinat, E.; Banci, L. A Unique Tool for Cellular Structural Biology: In-cell NMR. *J. Biol. Chem.* **2016**, *291*, 3776–3784. [CrossRef] [PubMed]
2. Luchinat, E.; Banci, L. In-cell NMR: A topical review. *IUCrJ* **2017**, *4 Pt 2*, 108–118. [CrossRef]
3. Luchinat, E.; Banci, L. In-Cell NMR in Human Cells: Direct Protein Expression Allows Structural Studies of Protein Folding and Maturation. *Acc. Chem. Res.* **2018**, *51*, 1550–1557. [CrossRef] [PubMed]
4. Kuhlbrandt, W. The resolution revolution. *Science* **2014**, *343*, 1443–1444. [CrossRef] [PubMed]
5. Fernandez-Leiro, R.; Scheres, S.H. Unravelling biological macromolecules with cryo-electron microscopy. *Nature* **2016**, *537*, 339–346. [CrossRef] [PubMed]
6. Giassa, I.C.; Rynes, J.; Fessl, T.; Foldynova-Trantirkova, S.; Trantirek, L. Advances in the cellular structural biology of nucleic acids. *FEBS Lett.* **2018**, *592*, 1997–2011. [CrossRef] [PubMed]
7. Irobalieva, R.N.; Martins, B.; Medalia, O. Cellular structural biology as revealed by cryo-electron tomography. *J. Cell Sci.* **2016**, *129*, 469–476. [CrossRef] [PubMed]
8. Ito, Y.; Selenko, P. Cellular structural biology. *Curr. Opin. Struct. Biol.* **2010**, *20*, 640–648. [CrossRef]
9. Koopmann, R.; Cupelli, K.; Redecke, L.; Nass, K.; Deponte, D.P.; White, T.A.; Stellato, F.; Rehders, D.; Liang, M.; Andreasson, J.; et al. In vivo protein crystallization opens new routes in structural biology. *Nat. Methods* **2012**, *9*, 259–262. [CrossRef]
10. Vinothkumar, K.R.; Henderson, R. Single particle electron cryomicroscopy: Trends, issues and future perspective. *Q. Rev. Biophys.* **2016**, *49*, e13. [CrossRef] [PubMed]
11. Danev, R.; Baumeister, W. Expanding the boundaries of cryo-EM with phase plates. *Curr. Opin. Struct. Biol.* **2017**, *46*, 87–94. [CrossRef] [PubMed]
12. Beck, M.; Baumeister, W. Cryo-Electron Tomography: Can it Reveal the Molecular Sociology of Cells in Atomic Detail? *Trends Cell Biol.* **2016**, *26*, 825–837. [CrossRef] [PubMed]

13. Hutchings, J.; Zanetti, G. Fine details in complex environments: The power of cryo-electron tomography. *Biochem. Soc. Trans.* **2018**, *46*, 807–816. [CrossRef] [PubMed]

14. Wagner, J.; Schaffer, M.; Fernandez-Busnadiego, R. Cryo-electron tomography-the cell biology that came in from the cold. *FEBS Lett.* **2017**, *591*, 2520–2533. [CrossRef] [PubMed]

15. Wolf, S.G.; Houben, L.; Elbaum, M. Cryo-scanning transmission electron tomography of vitrified cells. *Nat. Methods* **2014**, *11*, 423–428. [CrossRef] [PubMed]

16. Elbaum, M. Quantitative Cryo-Scanning Transmission Electron Microscopy of Biological Materials. *Adv. Mater.* **2018**, *30*, e1706681. [CrossRef] [PubMed]

17. De Boer, P.; Hoogenboom, J.P.; Giepmans, B.N. Correlated light and electron microscopy: Ultrastructure lights up! *Nat. Methods* **2015**, *12*, 503–513. [CrossRef] [PubMed]

18. Karreman, M.A.; Hyenne, V.; Schwab, Y.; Goetz, J.G. Intravital Correlative Microscopy: Imaging Life at the Nanoscale. *Trends Cell Biol.* **2016**, *26*, 848–863. [CrossRef]

19. Wolff, G.; Hagen, C.; Grunewald, K.; Kaufmann, R. Towards correlative super-resolution fluorescence and electron cryo-microscopy. *Biol. Cell* **2016**, *108*, 245–258. [CrossRef]

20. Oikonomou, C.M.; Jensen, G.J. Cellular Electron Cryotomography: Toward Structural Biology In situ. *Annu. Rev. Biochem.* **2017**, *86*, 873–896. [CrossRef]

21. Weber, M.S.; Wojtynek, M.; Medalia, O. Cellular and Structural Studies of Eukaryotic Cells by Cryo-Electron Tomography. *Cells* **2019**, *8*, 57. [CrossRef]

22. Briggs, J.A. Structural biology in situ—The potential of subtomogram averaging. *Curr. Opin. Struct. Biol.* **2013**, *23*, 261–267. [CrossRef]

23. Selenko, P.; Sprangers, R.; Stier, G.; Buhler, D.; Fischer, U.; Sattler, M. SMN tudor domain structure and its interaction with the Sm proteins. *Nat. Struct. Biol.* **2001**, *8*, 27–31.

24. Sprangers, R.; Selenko, P.; Sattler, M.; Sinning, I.; Groves, M.R. Definition of domain boundaries and crystallization of the SMN Tudor domain. *Acta Crystallogr. D Biol. Crystallogr.* **2003**, *59 Pt 2*, 366–368. [CrossRef]

25. Gronenborn, A.M. Harnessing the Combined Power of SAXS and NMR. *Adv. Exp. Med. Biol.* **2018**, *1105*, 171–180.

26. Pervushin, K. Impact of transverse relaxation optimized spectroscopy (TROSY) on NMR as a technique in structural biology. *Q. Rev. Biophys.* **2000**, *33*, 161–197. [CrossRef]

27. Wiesner, S.; Sprangers, R. Methyl groups as NMR probes for biomolecular interactions. *Curr. Opin. Struct. Biol.* **2015**, *35*, 60–67. [CrossRef]

28. Zhang, H.; van Ingen, H. Isotope-labeling strategies for solution NMR studies of macromolecular assemblies. *Curr. Opin. Struct. Biol.* **2016**, *38*, 75–82. [CrossRef]

29. Hiller, S.; Wider, G. Automated projection spectroscopy and its applications. *Top. Curr. Chem.* **2012**, *316*, 21–47.

30. Li, D.; Hansen, A.L.; Bruschweiler-Li, L.; Bruschweiler, R. Non-Uniform and Absolute Minimal Sampling for High-Throughput Multidimensional NMR Applications. *Chemistry* **2018**, *24*, 11535–11544. [CrossRef]

31. Wright, P.E.; Dyson, H.J. Intrinsically unstructured proteins: Re-assessing the protein structure-function paradigm. *J. Mol. Biol.* **1999**, *293*, 321–331. [CrossRef]

32. Dyson, H.J.; Wright, P.E. Intrinsically unstructured proteins and their functions. *Nat. Rev. Mol. Cell Biol.* **2005**, *6*, 197–208. [CrossRef]

33. Jensen, M.R.; Ruigrok, R.W.; Blackledge, M. Describing intrinsically disordered proteins at atomic resolution by NMR. *Curr. Opin. Struct. Biol.* **2013**, *23*, 426–435. [CrossRef]

34. Jensen, M.R.; Zweckstetter, M.; Huang, J.R.; Blackledge, M. Exploring free-energy landscapes of intrinsically disordered proteins at atomic resolution using NMR spectroscopy. *Chem. Rev.* **2014**, *114*, 6632–6660. [CrossRef]

35. Gibbs, E.B.; Cook, E.C.; Showalter, S.A. Application of NMR to studies of intrinsically disordered proteins. *Arch. Biochem. Biophys.* **2017**, *628*, 57–70. [CrossRef]

36. Schneider, R.; Blackledge, M.; Jensen, M.R. Elucidating binding mechanisms and dynamics of intrinsically disordered protein complexes using NMR spectroscopy. *Curr. Opin. Struct. Biol.* **2018**, *54*, 10–18. [CrossRef]

37. Milles, S.; Salvi, N.; Blackledge, M.; Jensen, M.R. Characterization of intrinsically disordered proteins and their dynamic complexes: From in vitro to cell-like environments. *Prog. Nucl. Magn. Reson. Spectrosc.* **2018**, *109*, 79–100. [CrossRef]

38. Shin, Y.; Brangwynne, C.P. Liquid phase condensation in cell physiology and disease. *Science* **2017**, *357*, 6357. [CrossRef]

39. Uversky, V.N. Intrinsically disordered proteins in overcrowded milieu: Membrane-less organelles, phase separation, and intrinsic disorder. *Curr. Opin. Struct. Biol.* **2017**, *44*, 18–30. [CrossRef]

40. Quinn, C.M.; Polenova, T. Structural biology of supramolecular assemblies by magic-angle spinning NMR spectroscopy. *Q. Rev. Biophys.* **2017**, *50*, e1. [CrossRef]

41. Meier, B.H.; Riek, R.; Bockmann, A. Emerging Structural Understanding of Amyloid Fibrils by Solid-State NMR. *Trends Biochem. Sci.* **2017**, *42*, 777–787. [CrossRef]

42. Opella, S.J.; Marassi, F.M. Applications of NMR to membrane proteins. *Arch. Biochem. Biophys.* **2017**, *628*, 92–101. [CrossRef]

43. Asami, S.; Reif, B. Proton-detected solid-state NMR spectroscopy at aliphatic sites: Application to crystalline systems. *Acc. Chem. Res.* **2013**, *46*, 2089–2097. [CrossRef]

44. Jaudzems, K.; Polenova, T.; Pintacuda, G.; Oschkinat, H.; Lesage, A. DNP NMR of biomolecular assemblies. *J. Struct. Biol.* **2018**. [CrossRef]

45. Ritter, C.; Maddelein, M.L.; Siemer, A.B.; Luhrs, T.; Ernst, M.; Meier, B.H.; Saupe, S.J.; Riek, R. Correlation of structural elements and infectivity of the HET-s prion. *Nature* **2005**, *435*, 844–848. [CrossRef]

46. Wasmer, C.; Lange, A.; Van Melckebeke, H.; Siemer, A.B.; Riek, R.; Meier, B.H. Amyloid fibrils of the HET-s(218–289) prion form a beta solenoid with a triangular hydrophobic core. *Science* **2008**, *319*, 1523–1526. [CrossRef]

47. Van Melckebeke, H.; Wasmer, C.; Lange, A.; Ab, E.; Loquet, A.; Bockmann, A.; Meier, B.H. Atomic-resolution three-dimensional structure of HET-s(218–289) amyloid fibrils by solid-state NMR spectroscopy. *J. Am. Chem. Soc.* **2010**, *132*, 13765–13775. [CrossRef]

48. Chen, B.; Thurber, K.R.; Shewmaker, F.; Wickner, R.B.; Tycko, R. Measurement of amyloid fibril mass-per-length by tilted-beam transmission electron microscopy. *Proc. Natl. Acad. Sci. USA* **2009**, *106*, 14339–14344. [CrossRef]

49. Mizuno, N.; Baxa, U.; Steven, A.C. Structural dependence of HET-s amyloid fibril infectivity assessed by cryoelectron microscopy. *Proc. Natl. Acad. Sci. USA* **2011**, *108*, 3252–3257. [CrossRef]

50. Walti, M.A.; Ravotti, F.; Arai, H.; Glabe, C.G.; Wall, J.S.; Bockmann, A.; Guntert, P.; Meier, B.H.; Riek, R. Atomic-resolution structure of a disease-relevant Abeta(1-42) amyloid fibril. *Proc. Natl. Acad. Sci. USA* **2016**, *113*, E4976–E4984. [CrossRef]

51. Qiang, W.; Yau, W.M.; Lu, J.X.; Collinge, J.; Tycko, R. Structural variation in amyloid-beta fibrils from Alzheimer's disease clinical subtypes. *Nature* **2017**, *541*, 217–221. [CrossRef]

52. Schmidt, M.; Rohou, A.; Lasker, K.; Yadav, J.K.; Schiene-Fischer, C.; Fandrich, M.; Grigorieff, N. Peptide dimer structure in an Abeta(1-42) fibril visualized with cryo-EM. *Proc. Natl. Acad. Sci. USA* **2015**, *112*, 11858–11863. [CrossRef]

53. Gremer, L.; Scholzel, D.; Schenk, C.; Reinartz, E.; Labahn, J.; Ravelli, R.B.G.; Tusche, M.; Lopez-Iglesias, C.; Hoyer, W.; Heise, H.; et al. Fibril structure of amyloid-beta(1-42) by cryo-electron microscopy. *Science* **2017**, *358*, 116–119. [CrossRef]

54. Heise, H.; Hoyer, W.; Becker, S.; Andronesi, O.C.; Riedel, D.; Baldus, M. Molecular-level secondary structure, polymorphism, and dynamics of full-length alpha-synuclein fibrils studied by solid-state NMR. *Proc. Natl. Acad. Sci. USA* **2005**, *102*, 15871–15876. [CrossRef]

55. Heise, H.; Celej, M.S.; Becker, S.; Riedel, D.; Pelah, A.; Kumar, A.; Jovin, T.M.; Baldus, M. Solid-state NMR reveals structural differences between fibrils of wild-type and disease-related A53T mutant alpha-synuclein. *J. Mol. Biol.* **2008**, *380*, 444–450. [CrossRef]

56. Leftin, A.; Job, C.; Beyer, K.; Brown, M.F. Solid-state (1)(3)C NMR reveals annealing of raft-like membranes containing cholesterol by the intrinsically disordered protein alpha-Synuclein. *J. Mol. Biol.* **2013**, *425*, 2973–2987. [CrossRef]

57. Villa, E.; Schaffer, M.; Plitzko, J.M.; Baumeister, W. Opening windows into the cell: Focused-ion-beam milling for cryo-electron tomography. *Curr. Opin. Struct. Biol.* **2013**, *23*, 771–777. [CrossRef]

58. Gath, J.; Bousset, L.; Habenstein, B.; Melki, R.; Bockmann, A.; Meier, B.H. Unlike twins: An NMR comparison of two alpha-synuclein polymorphs featuring different toxicity. *PLoS ONE* **2014**, *9*, e90659. [CrossRef]

59. Tuttle, M.D.; Comellas, G.; Nieuwkoop, A.J.; Covell, D.J.; Berthold, D.A.; Kloepper, K.D.; Courtney, J.M.; Kim, J.K.; Barclay, A.M.; Kendall, A.; et al. Solid-state NMR structure of a pathogenic fibril of full-length human alpha-synuclein. *Nat. Struct. Mol. Biol.* **2016**, *23*, 409–415. [CrossRef]

60. Hwang, S.; Fricke, P.; Zinke, M.; Giller, K.; Wall, J.S.; Riedel, D.; Becker, S.; Lange, A. Comparison of the 3D structures of mouse and human alpha-synuclein fibrils by solid-state NMR and STEM. *J. Struct. Biol.* **2018**. [CrossRef]

61. Vilar, M.; Chou, H.T.; Luhrs, T.; Maji, S.K.; Riek-Loher, D.; Verel, R.; Manning, G.; Stahlberg, H.; Riek, R. The fold of alpha-synuclein fibrils. *Proc. Natl. Acad. Sci. USA* **2008**, *105*, 8637–8642. [CrossRef]

62. Li, B.; Ge, P.; Murray, K.A.; Sheth, P.; Zhang, M.; Nair, G.; Sawaya, M.R.; Shin, W.S.; Boyer, D.R.; Ye, S.; et al. Cryo-EM of full-length alpha-synuclein reveals fibril polymorphs with a common structural kernel. *Nat. Commun.* **2018**, *9*, 3609. [CrossRef]

63. Andronesi, O.C.; von Bergen, M.; Biernat, J.; Seidel, K.; Griesinger, C.; Mandelkow, E.; Baldus, M. Characterization of Alzheimer's-like paired helical filaments from the core domain of tau protein using solid-state NMR spectroscopy. *J. Am. Chem. Soc.* **2008**, *130*, 5922–5928. [CrossRef]

64. Daebel, V.; Chinnathambi, S.; Biernat, J.; Schwalbe, M.; Habenstein, B.; Loquet, A.; Akoury, E.; Tepper, K.; Muller, H.; Baldus, M.; et al. beta-Sheet core of tau paired helical filaments revealed by solid-state NMR. *J. Am. Chem. Soc.* **2012**, *134*, 13982–13989. [CrossRef]

65. Xiang, S.; Kulminskaya, N.; Habenstein, B.; Biernat, J.; Tepper, K.; Paulat, M.; Griesinger, C.; Becker, S.; Lange, A.; Mandelkow, E.; et al. A Two-Component Adhesive: Tau Fibrils Arise from a Combination of a Well-Defined Motif and Conformationally Flexible Interactions. *J. Am. Chem. Soc.* **2017**, *139*, 2639–2646. [CrossRef]

66. Falcon, B.; Zhang, W.; Murzin, A.G.; Murshudov, G.; Garringer, H.J.; Vidal, R.; Crowther, R.A.; Ghetti, B.; Scheres, S.H.W.; Goedert, M. Structures of filaments from Pick's disease reveal a novel tau protein fold. *Nature* **2018**, *561*, 137–140. [CrossRef]

67. Falcon, B.; Zhang, W.; Schweighauser, M.; Murzin, A.G.; Vidal, R.; Garringer, H.J.; Ghetti, B.; Scheres, S.H.W.; Goedert, M. Tau filaments from multiple cases of sporadic and inherited Alzheimer's disease adopt a common fold. *Acta Neuropathol.* **2018**, *136*, 699–708. [CrossRef]

68. Fitzpatrick, A.W.P.; Falcon, B.; He, S.; Murzin, A.G.; Murshudov, G.; Garringer, H.J.; Crowther, R.A.; Ghetti, B.; Goedert, M.; Scheres, S.H.W. Cryo-EM structures of tau filaments from Alzheimer's disease. *Nature* **2017**, *547*, 185–190. [CrossRef]

69. Bayro, M.J.; Ganser-Pornillos, B.K.; Zadrozny, K.K.; Yeager, M.; Tycko, R. Helical Conformation in the CA-SP1 Junction of the Immature HIV-1 Lattice Determined from Solid-State NMR of Virus-like Particles. *J. Am. Chem. Soc.* **2016**, *138*, 12029–12032. [CrossRef]

70. Bayro, M.J.; Tycko, R. Structure of the Dimerization Interface in the Mature HIV-1 Capsid Protein Lattice from Solid State NMR of Tubular Assemblies. *J. Am. Chem. Soc.* **2016**, *138*, 8538–8546. [CrossRef]

71. Byeon, I.J.; Meng, X.; Jung, J.; Zhao, G.; Yang, R.; Ahn, J.; Shi, J.; Concel, J.; Aiken, C.; Zhang, P.; et al. Structural convergence between Cryo-EM and NMR reveals intersubunit interactions critical for HIV-1 capsid function. *Cell* **2009**, *139*, 780–790. [CrossRef]

72. Demers, J.P.; Habenstein, B.; Loquet, A.; Kumar Vasa, S.; Giller, K.; Becker, S.; Baker, D.; Lange, A.; Sgourakis, N.G. High-resolution structure of the Shigella type-III secretion needle by solid-state NMR and cryo-electron microscopy. *Nat. Commun.* **2014**, *5*, 4976. [CrossRef]

73. Loquet, A.; Sgourakis, N.G.; Gupta, R.; Giller, K.; Riedel, D.; Goosmann, C.; Griesinger, C.; Kolbe, M.; Baker, D.; Becker, S.; et al. Atomic model of the type III secretion system needle. *Nature* **2012**, *486*, 276–279. [CrossRef]

74. Fujii, T.; Cheung, M.; Blanco, A.; Kato, T.; Blocker, A.J.; Namba, K. Structure of a type III secretion needle at 7-A resolution provides insights into its assembly and signaling mechanisms. *Proc. Natl. Acad. Sci. USA* **2012**, *109*, 4461–4466. [CrossRef]

75. Hu, J.; Worrall, L.J.; Hong, C.; Vuckovic, M.; Atkinson, C.E.; Caveney, N.; Yu, Z.; Strynadka, N.C.J. Cryo-EM analysis of the T3S injectisome reveals the structure of the needle and open secretin. *Nat. Commun.* **2018**, *9*, 3840. [CrossRef]

76. Worrall, L.J.; Hong, C.; Vuckovic, M.; Deng, W.; Bergeron, J.R.; Majewski, D.D.; Huang, R.K.; Spreter, T.; Finlay, B.B.; Yu, Z.; et al. Near-atomic-resolution cryo-EM analysis of the Salmonella T3S injectisome basal body. *Nature* **2016**, *540*, 597–601. [CrossRef]

77. Pinto, C.; Mance, D.; Sinnige, T.; Daniels, M.; Weingarth, M.; Baldus, M. Formation of the beta-barrel assembly machinery complex in lipid bilayers as seen by solid-state NMR. *Nat. Commun.* **2018**, *9*, 4135. [CrossRef]

78. Retel, J.S.; Nieuwkoop, A.J.; Hiller, M.; Higman, V.A.; Barbet-Massin, E.; Stanek, J.; Andreas, L.B.; Franks, W.T.; van Rossum, B.J.; Vinothkumar, K.R.; et al. Structure of outer membrane protein G in lipid bilayers. *Nat. Commun.* **2017**, *8*, 2073. [CrossRef]

79. Iadanza, M.G.; Higgins, A.J.; Schiffrin, B.; Calabrese, A.N.; Brockwell, D.J.; Ashcroft, A.E.; Radford, S.E.; Ranson, N.A. Lateral opening in the intact beta-barrel assembly machinery captured by cryo-EM. *Nat. Commun.* **2016**, *7*, 12865. [CrossRef]

80. Nans, A.; Kudryashev, M.; Saibil, H.R.; Hayward, R.D. Structure of a bacterial type III secretion system in contact with a host membrane in situ. *Nat. Commun.* **2015**, *6*, 10114. [CrossRef]

81. Hu, B.; Lara-Tejero, M.; Kong, Q.; Galan, J.E.; Liu, J. In situ Molecular Architecture of the Salmonella Type III Secretion Machine. *Cell* **2017**, *168*, 1065.e10–1074.e10. [CrossRef]

82. Serber, Z.; Keatinge-Clay, A.T.; Ledwidge, R.; Kelly, A.E.; Miller, S.M.; Dotsch, V. High-resolution macromolecular NMR spectroscopy inside living cells. *J. Am. Chem. Soc.* **2001**, *123*, 2446–2447. [CrossRef]

83. Barbieri, L.; Luchinat, E.; Banci, L. Characterization of proteins by in-cell NMR spectroscopy in cultured mammalian cells. *Nat. Protoc.* **2016**, *11*, 1101–1111. [CrossRef]

84. Bekei, B.; Rose, H.M.; Herzig, M.; Dose, A.; Schwarzer, D.; Selenko, P. In-cell NMR in mammalian cells: Part 1. *Methods Mol. Biol.* **2012**, *895*, 43–54.

85. Bekei, B.; Rose, H.M.; Herzig, M.; Selenko, P. In-cell NMR in mammalian cells: Part 2. *Methods Mol. Biol.* **2012**, *895*, 55–66.

86. Bekei, B.; Rose, H.M.; Herzig, M.; Stephanowitz, H.; Krause, E.; Selenko, P. In-cell NMR in mammalian cells: Part 3. *Methods Mol. Biol.* **2012**, *895*, 67–83.

87. Walsh, C.T.; Garneau-Tsodikova, S.; Gatto, G.J., Jr. Protein posttranslational modifications: The chemistry of proteome diversifications. *Angew. Chem. Int. Ed. Engl.* **2005**, *44*, 7342–7372. [CrossRef]

88. Dunker, A.K.; Uversky, V.N. Signal transduction via unstructured protein conduits. *Nat. Chem. Biol.* **2008**, *4*, 229–230. [CrossRef]

89. Wright, P.E.; Dyson, H.J. Intrinsically disordered proteins in cellular signalling and regulation. *Nat. Rev. Mol. Cell. Biol.* **2015**, *16*, 18–29. [CrossRef]

90. Csizmok, V.; Forman-Kay, J.D. Complex regulatory mechanisms mediated by the interplay of multiple post-translational modifications. *Curr. Opin. Struct. Biol.* **2018**, *48*, 58–67. [CrossRef]

91. Theillet, F.X.; Smet-Nocca, C.; Liokatis, S.; Thongwichian, R.; Kosten, J.; Yoon, M.K.; Kriwacki, R.W.; Landrieu, I.; Lippens, G.; Selenko, P. Cell signaling, post-translational protein modifications and NMR spectroscopy. *J. Biomol. NMR* **2012**, *54*, 217–236. [CrossRef]

92. Selenko, P.; Frueh, D.P.; Elsaesser, S.J.; Haas, W.; Gygi, S.P.; Wagner, G. In situ observation of protein phosphorylation by high-resolution NMR spectroscopy. *Nat. Struct. Mol. Biol.* **2008**, *15*, 321–329. [CrossRef]

93. Baker, J.M.; Hudson, R.P.; Kanelis, V.; Choy, W.Y.; Thibodeau, P.H.; Thomas, P.J.; Forman-Kay, J.D. CFTR regulatory region interacts with NBD1 predominantly via multiple transient helices. *Nat. Struct. Mol. Biol.* **2007**, *14*, 738–745. [CrossRef]

94. Byeon, I.J.; Li, H.; Song, H.; Gronenborn, A.M.; Tsai, M.D. Sequential phosphorylation and multisite interactions characterize specific target recognition by the FHA domain of Ki67. *Nat. Struct. Mol. Biol.* **2005**, *12*, 987–993. [CrossRef]

95. Landrieu, I.; Lacosse, L.; Leroy, A.; Wieruszeski, J.M.; Trivelli, X.; Sillen, A.; Sibille, N.; Schwalbe, H.; Saxena, K.; Langer, T.; et al. NMR analysis of a Tau phosphorylation pattern. *J. Am. Chem. Soc.* **2006**, *128*, 3575–3583. [CrossRef]

96. Kumar, G.S.; Page, R.; Peti, W. Preparation of Phosphorylated Proteins for NMR Spectroscopy. *Methods Enzymol.* **2019**, *614*, 187–205.

97. Smith, M.J.; Marshall, C.B.; Theillet, F.X.; Binolfi, A.; Selenko, P.; Ikura, M. Real-time NMR monitoring of biological activities in complex physiological environments. *Curr. Opin. Struct. Biol.* **2015**, *32*, 39–47. [CrossRef]

98. Theillet, F.X.; Rose, H.M.; Liokatis, S.; Binolfi, A.; Thongwichian, R.; Stuiver, M.; Selenko, P. Site-specific NMR mapping and time-resolved monitoring of serine and threonine phosphorylation in reconstituted kinase reactions and mammalian cell extracts. *Nat. Protoc.* **2013**, *8*, 1416–1432. [CrossRef]

99. Amata, I.; Maffei, M.; Igea, A.; Gay, M.; Vilaseca, M.; Nebreda, A.R.; Pons, M. Multi-phosphorylation of the intrinsically disordered unique domain of c-Src studied by in-cell and real-time NMR spectroscopy. *ChemBioChem* **2013**, *14*, 1820–1827. [CrossRef]

100. Bachman, A.B.; Keramisanou, D.; Xu, W.; Beebe, K.; Moses, M.A.; Vasantha Kumar, M.V.; Gray, G.; Noor, R.E.; van der Vaart, A.; Neckers, L.; et al. Phosphorylation induced cochaperone unfolding promotes kinase recruitment and client class-specific Hsp90 phosphorylation. *Nat. Commun.* **2018**, *9*, 265. [CrossRef]

101. Bozoky, Z.; Ahmadi, S.; Milman, T.; Kim, T.H.; Du, K.; Di Paola, M.; Pasyk, S.; Pekhletski, R.; Keller, J.P.; Bear, C.E.; et al. Synergy of cAMP and calcium signaling pathways in CFTR regulation. *Proc. Natl. Acad. Sci. USA* **2017**, *114*, E2086–E2095. [CrossRef] [PubMed]

102. Breindel, L.; DeMott, C.; Burz, D.S.; Shekhtman, A. Real-Time In-Cell Nuclear Magnetic Resonance: Ribosome-Targeted Antibiotics Modulate Quinary Protein Interactions. *Biochemistry* **2018**, *57*, 540–546. [CrossRef] [PubMed]

103. Cordier, F.; Chaffotte, A.; Terrien, E.; Prehaud, C.; Theillet, F.X.; Delepierre, M.; Lafon, M.; Buc, H.; Wolff, N. Ordered phosphorylation events in two independent cascades of the PTEN C-tail revealed by NMR. *J. Am. Chem. Soc.* **2012**, *134*, 20533–20543. [CrossRef]

104. Despres, C.; Byrne, C.; Qi, H.; Cantrelle, F.X.; Huvent, I.; Chambraud, B.; Baulieu, E.E.; Jacquot, Y.; Landrieu, I.; Lippens, G.; et al. Identification of the Tau phosphorylation pattern that drives its aggregation. *Proc. Natl. Acad. Sci. USA* **2017**, *114*, 9080–9085. [CrossRef] [PubMed]

105. Gibbs, E.B.; Lu, F.; Portz, B.; Fisher, M.J.; Medellin, B.P.; Laremore, T.N.; Zhang, Y.J.; Gilmour, D.S.; Showalter, S.A. Phosphorylation induces sequence-specific conformational switches in the RNA polymerase II C-terminal domain. *Nat. Commun.* **2017**, *8*, 15233. [CrossRef] [PubMed]

106. Gladkova, C.; Schubert, A.F.; Wagstaff, J.L.; Pruneda, J.N.; Freund, S.M.; Komander, D. An invisible ubiquitin conformation is required for efficient phosphorylation by PINK1. *EMBO J.* **2017**, *36*, 3555–3572. [CrossRef]

107. Guca, E.; Sunol, D.; Ruiz, L.; Konkol, A.; Cordero, J.; Torner, C.; Aragon, E.; Martin-Malpartida, P.; Riera, A.; Macias, M.J. TGIF1 homeodomain interacts with Smad MH1 domain and represses TGF-beta signaling. *Nucleic Acids Res.* **2018**, *46*, 9220–9235. [CrossRef]

108. Hendus-Altenburger, R.; Lambrughi, M.; Terkelsen, T.; Pedersen, S.F.; Papaleo, E.; Lindorff-Larsen, K.; Kragelund, B.B. A phosphorylation-motif for tuneable helix stabilisation in intrinsically disordered proteins—Lessons from the sodium proton exchanger 1 (NHE1). *Cell. Signal.* **2017**, *37*, 40–51. [CrossRef]

109. Hendus-Altenburger, R.; Pedraz-Cuesta, E.; Olesen, C.W.; Papaleo, E.; Schnell, J.A.; Hopper, J.T.; Robinson, C.V.; Pedersen, S.F.; Kragelund, B.B. The human Na(+)/H(+) exchanger 1 is a membrane scaffold protein for extracellular signal-regulated kinase 2. *BMC Biol.* **2016**, *14*, 31. [CrossRef]

110. Himmel, S.; Zschiedrich, C.P.; Becker, S.; Hsiao, H.H.; Wolff, S.; Diethmaier, C.; Urlaub, H.; Lee, D.; Griesinger, C.; Stulke, J. Determinants of interaction specificity of the Bacillus subtilis GlcT antitermination protein: Functionality and phosphorylation specificity depend on the arrangement of the regulatory domains. *J. Biol. Chem.* **2012**, *287*, 27731–27742. [CrossRef]

111. Kano, Y.; Gebregiworgis, T.; Marshall, C.B.; Radulovich, N.; Poon, B.P.K.; St-Germain, J.; Cook, J.D.; Valencia-Sama, I.; Grant, B.M.M.; Herrera, S.G.; et al. Tyrosyl phosphorylation of KRAS stalls GTPase cycle via alteration of switch I and II conformation. *Nat. Commun.* **2019**, *10*, 224. [CrossRef] [PubMed]

112. Kosten, J.; Binolfi, A.; Stuiver, M.; Verzini, S.; Theillet, F.X.; Bekei, B.; van Rossum, M.; Selenko, P. Efficient modification of alpha-synuclein serine 129 by protein kinase CK1 requires phosphorylation of tyrosine 125 as a priming event. *ACS Chem. Neurosci.* **2014**, *5*, 1203–1208. [CrossRef] [PubMed]

113. Kulkarni, P.; Jolly, M.K.; Jia, D.; Mooney, S.M.; Bhargava, A.; Kagohara, L.T.; Chen, Y.; Hao, P.; He, Y.; Veltri, R.W.; et al. Phosphorylation-induced conformational dynamics in an intrinsically disordered protein and potential role in phenotypic heterogeneity. *Proc. Natl. Acad. Sci. USA* **2017**, *114*, E2644–E2653. [CrossRef] [PubMed]

114. Kumar, A.; Gopalswamy, M.; Wishart, C.; Henze, M.; Eschen-Lippold, L.; Donnelly, D.; Balbach, J. N-terminal phosphorylation of parathyroid hormone (PTH) abolishes its receptor activity. *ACS Chem. Biol.* **2014**, *9*, 2465–2470. [CrossRef]

115. Liokatis, S.; Klingberg, R.; Tan, S.; Schwarzer, D. Differentially Isotope-Labeled Nucleosomes To Study Asymmetric Histone Modification Crosstalk by Time-Resolved NMR Spectroscopy. *Angew. Chem. Int. Ed. Engl.* **2016**, *55*, 8262–8265. [CrossRef] [PubMed]

116. Lousa, P.; Nedozralova, H.; Zupa, E.; Novacek, J.; Hritz, J. Phosphorylation of the regulatory domain of human tyrosine hydroxylase 1 monitored using non-uniformly sampled NMR. *Biophys. Chem.* **2017**, *223*, 25–29. [CrossRef] [PubMed]

117. Mayzel, M.; Rosenlow, J.; Isaksson, L.; Orekhov, V.Y. Time-resolved multidimensional NMR with non-uniform sampling. *J. Biomol. NMR* **2014**, *58*, 129–139. [CrossRef]

118. Mbefo, M.K.; Fares, M.B.; Paleologou, K.; Oueslati, A.; Yin, G.; Tenreiro, S.; Pinto, M.; Outeiro, T.; Zweckstetter, M.; Masliah, E.; et al. Parkinson disease mutant E46K enhances alpha-synuclein phosphorylation in mammalian cell lines, in yeast, and in vivo. *J. Biol. Chem.* **2015**, *290*, 9412–9427. [CrossRef] [PubMed]

119. Mbefo, M.K.; Paleologou, K.E.; Boucharaba, A.; Oueslati, A.; Schell, H.; Fournier, M.; Olschewski, D.; Yin, G.; Zweckstetter, M.; Masliah, E.; et al. Phosphorylation of synucleins by members of the Polo-like kinase family. *J. Biol. Chem.* **2010**, *285*, 2807–2822. [CrossRef]

120. Munari, F.; Gajda, M.J.; Hiragami-Hamada, K.; Fischle, W.; Zweckstetter, M. Characterization of the effects of phosphorylation by CK2 on the structure and binding properties of human HP1beta. *FEBS Lett.* **2014**, *588*, 1094–1099. [CrossRef]

121. Mylona, A.; Theillet, F.X.; Foster, C.; Cheng, T.M.; Miralles, F.; Bates, P.A.; Selenko, P.; Treisman, R. Opposing effects of Elk-1 multisite phosphorylation shape its response to ERK activation. *Science* **2016**, *354*, 233–237. [CrossRef]

122. Narasimamurthy, R.; Hunt, S.R.; Lu, Y.; Fustin, J.M.; Okamura, H.; Partch, C.L.; Forger, D.B.; Kim, J.K.; Virshup, D.M. CK1delta/epsilon protein kinase primes the PER2 circadian phosphoswitch. *Proc. Natl. Acad. Sci. USA* **2018**, *115*, 5986–5991. [CrossRef]

123. Nogueira, M.O.; Hosek, T.; Calcada, E.O.; Castiglia, F.; Massimi, P.; Banks, L.; Felli, I.C.; Pierattelli, R. Monitoring HPV-16 E7 phosphorylation events. *Virology* **2017**, *503*, 70–75. [CrossRef]

124. Okuda, M.; Nishimura, Y. Real-time and simultaneous monitoring of the phosphorylation and enhanced interaction of p53 and XPC acidic domains with the TFIIH p62 subunit. *Oncogenesis* **2015**, *4*, e150. [CrossRef]

125. Peterson, D.W.; Ando, D.M.; Taketa, D.A.; Zhou, H.; Dahlquist, F.W.; Lew, J. No difference in kinetics of tau or histone phosphorylation by CDK5/p25 versus CDK5/p35 in vitro. *Proc. Natl. Acad. Sci. USA* **2010**, *107*, 2884–2889. [CrossRef]

126. Qi, H.; Prabakaran, S.; Cantrelle, F.X.; Chambraud, B.; Gunawardena, J.; Lippens, G.; Landrieu, I. Characterization of Neuronal Tau Protein as a Target of Extracellular Signal-regulated Kinase. *J. Biol. Chem.* **2016**, *291*, 7742–7753. [CrossRef]

127. Rose, H.M.; Stuiver, M.; Thongwichian, R.; Theillet, F.X.; Feller, S.M.; Selenko, P. Quantitative NMR analysis of Erk activity and inhibition by U0126 in a panel of patient-derived colorectal cancer cell lines. *Biochim. Biophys. Acta* **2013**, *1834*, 1396–1401. [CrossRef]

128. Rosenlow, J.; Isaksson, L.; Mayzel, M.; Lengqvist, J.; Orekhov, V.Y. Tyrosine phosphorylation within the intrinsically disordered cytosolic domains of the B-cell receptor: An NMR-based structural analysis. *PLoS ONE* **2014**, *9*, e96199. [CrossRef]

129. Schwalbe, M.; Biernat, J.; Bibow, S.; Ozenne, V.; Jensen, M.R.; Kadavath, H.; Blackledge, M.; Mandelkow, E.; Zweckstetter, M. Phosphorylation of human Tau protein by microtubule affinity-regulating kinase 2. *Biochemistry* **2013**, *52*, 9068–9079. [CrossRef]

130. Secci, E.; Luchinat, E.; Banci, L. The Casein Kinase 2-Dependent Phosphorylation of NS5A Domain 3 from Hepatitis C Virus Followed by Time-Resolved NMR Spectroscopy. *ChemBioChem* **2016**, *17*, 328–333. [CrossRef]

131. Smet-Nocca, C.; Launay, H.; Wieruszeski, J.M.; Lippens, G.; Landrieu, I. Unraveling a phosphorylation event in a folded protein by NMR spectroscopy: Phosphorylation of the Pin1 WW domain by PKA. *J. Biomol. NMR* **2013**, *55*, 323–337. [CrossRef]

132. Solyom, Z.; Ma, P.; Schwarten, M.; Bosco, M.; Polidori, A.; Durand, G.; Willbold, D.; Brutscher, B. The Disordered Region of the HCV Protein NS5A: Conformational Dynamics, SH3 Binding, and Phosphorylation. *Biophys. J.* **2015**, *109*, 1483–1496. [CrossRef]

133. Stott, K.; Watson, M.; Bostock, M.J.; Mortensen, S.A.; Travers, A.; Grasser, K.D.; Thomas, J.O. Structural insights into the mechanism of negative regulation of single-box high mobility group proteins by the acidic tail domain. *J. Biol. Chem.* **2014**, *289*, 29817–29826. [CrossRef]

134. Stutzer, A.; Liokatis, S.; Kiesel, A.; Schwarzer, D.; Sprangers, R.; Soding, J.; Selenko, P.; Fischle, W. Modulations of DNA Contacts by Linker Histones and Post-translational Modifications Determine the Mobility and Modifiability of Nucleosomal H3 Tails. *Mol. Cell* **2016**, *61*, 247–259. [CrossRef]

135. Thongwichian, R.; Kosten, J.; Benary, U.; Rose, H.M.; Stuiver, M.; Theillet, F.X.; Dose, A.; Koch, B.; Yokoyama, H.; Schwarzer, D.; et al. A Multiplexed NMR-Reporter Approach to Measure Cellular Kinase and Phosphatase Activities in Real-Time. *J. Am. Chem. Soc.* **2015**, *137*, 6468–6471. [CrossRef]

136. Liokatis, S.; Dose, A.; Schwarzer, D.; Selenko, P. Simultaneous detection of protein phosphorylation and acetylation by high-resolution NMR spectroscopy. *J. Am. Chem. Soc.* **2010**, *132*, 14704–14705. [CrossRef]

137. Smet-Nocca, C.; Page, A.; Cantrelle, F.X.; Nikolakaki, E.; Landrieu, I.; Giannakouros, T. The O-beta-linked N-acetylglucosaminylation of the Lamin B receptor and its impact on DNA binding and phosphorylation. *Biochim. Biophys. Acta Gen. Subj.* **2018**, *1862*, 825–835. [CrossRef]

138. Theillet, F.X.; Liokatis, S.; Jost, J.O.; Bekei, B.; Rose, H.M.; Binolfi, A.; Schwarzer, D.; Selenko, P. Site-specific mapping and time-resolved monitoring of lysine methylation by high-resolution NMR spectroscopy. *J. Am. Chem. Soc.* **2012**, *134*, 7616–7619. [CrossRef]

139. Binolfi, A.; Limatola, A.; Verzini, S.; Kosten, J.; Theillet, F.X.; Rose, H.M.; Bekei, B.; Stuiver, M.; van Rossum, M.; Selenko, P. Intracellular repair of oxidation-damaged alpha-synuclein fails to target C-terminal modification sites. *Nat. Commun.* **2016**, *7*, 10251. [CrossRef]

140. Limatola, A.; Eichmann, C.; Jacob, R.S.; Ben-Nissan, G.; Sharon, M.; Binolfi, A.; Selenko, P. Time-Resolved NMR Analysis of Proteolytic alpha-Synuclein Processing in vitro and *in cellulo*. *Proteomics* **2018**, *18*, e1800056. [CrossRef]

141. Johnson, L.N.; Barford, D. The effects of phosphorylation on the structure and function of proteins. *Annu. Rev. Biophys. Biomol. Struct.* **1993**, *22*, 199–232. [CrossRef] [PubMed]

142. Elbaum, M.B.; Zondlo, N.J. OGlcNAcylation and phosphorylation have similar structural effects in alpha-helices: Post-translational modifications as inducible start and stop signals in alpha-helices, with greater structural effects on threonine modification. *Biochemistry* **2014**, *53*, 2242–2260. [CrossRef] [PubMed]

143. Andrew, C.D.; Warwicker, J.; Jones, G.R.; Doig, A.J. Effect of phosphorylation on alpha-helix stability as a function of position. *Biochemistry* **2002**, *41*, 1897–1905. [CrossRef]

144. Johnson, L.N.; Lewis, R.J. Structural basis for control by phosphorylation. *Chem. Rev.* **2001**, *101*, 2209–2242. [CrossRef] [PubMed]

145. Macek, P.; Cliff, M.J.; Embrey, K.J.; Holdgate, G.A.; Nissink, J.W.M.; Panova, S.; Waltho, J.P.; Davies, R.A. Myc phosphorylation in its basic helix-loop-helix region destabilizes transient alpha-helical structures, disrupting Max and DNA binding. *J. Biol. Chem.* **2018**, *293*, 9301–9310. [CrossRef] [PubMed]

146. Kim, D.H.; Han, K.H. Transient Secondary Structures as General Target-Binding Motifs in Intrinsically Disordered Proteins. *Int. J. Mol. Sci.* **2018**, *19*, 3614. [CrossRef]

147. Wright, P.E.; Dyson, H.J. Linking folding and binding. *Curr. Opin. Struct. Biol.* **2009**, *19*, 31–38. [CrossRef]

148. Darling, A.L.; Uversky, V.N. Intrinsic Disorder and Posttranslational Modifications: The Darker Side of the Biological Dark Matter. *Front. Genet.* **2018**, *9*, 158. [CrossRef]

149. Condos, T.E.; Dunkerley, K.M.; Freeman, E.A.; Barber, K.R.; Aguirre, J.D.; Chaugule, V.K.; Xiao, Y.; Konermann, L.; Walden, H.; Shaw, G.S. Synergistic recruitment of UbcH7~Ub and phosphorylated Ubl domain triggers parkin activation. *EMBO J.* **2018**, *37*, e100014. [CrossRef]

150. Kumar, G.S.; Clarkson, M.W.; Kunze, M.B.A.; Granata, D.; Wand, A.J.; Lindorff-Larsen, K.; Page, R.; Peti, W. Dynamic activation and regulation of the mitogen-activated protein kinase p38. *Proc. Natl. Acad. Sci. USA* **2018**, *115*, 4655–4660. [CrossRef]

151. Martin, E.W.; Holehouse, A.S.; Grace, C.R.; Hughes, A.; Pappu, R.V.; Mittag, T. Sequence Determinants of the Conformational Properties of an Intrinsically Disordered Protein Prior to and upon Multisite Phosphorylation. *J. Am. Chem. Soc.* **2016**, *138*, 15323–15335. [CrossRef] [PubMed]

152. Monahan, Z.; Ryan, V.H.; Janke, A.M.; Burke, K.A.; Rhoads, S.N.; Zerze, G.H.; O'Meally, R.; Dignon, G.L.; Conicella, A.E.; Zheng, W.; et al. Phosphorylation of the FUS low-complexity domain disrupts phase separation, aggregation, and toxicity. *EMBO J.* **2017**, *36*, 2951–2967. [CrossRef]

153. Patel, P.; Prescott, G.R.; Burgoyne, R.D.; Lian, L.Y.; Morgan, A. Phosphorylation of Cysteine String Protein Triggers a Major Conformational Switch. *Structure* **2016**, *24*, 1380–1386. [CrossRef] [PubMed]

154. Perez-Borrajero, C.; Lin, C.S.; Okon, M.; Scheu, K.; Graves, B.J.; Murphy, M.E.P.; McIntosh, L.P. The Biophysical Basis for Phosphorylation-Enhanced DNA-Binding Autoinhibition of the ETS1 Transcription Factor. *J. Mol. Biol.* **2019**, *431*, 593–614. [CrossRef] [PubMed]

155. Schwalbe, M.; Kadavath, H.; Biernat, J.; Ozenne, V.; Blackledge, M.; Mandelkow, E.; Zweckstetter, M. Structural Impact of Tau Phosphorylation at Threonine 231. *Structure* **2015**, *23*, 1448–1458. [CrossRef] [PubMed]

156. Shiraishi, Y.; Natsume, M.; Kofuku, Y.; Imai, S.; Nakata, K.; Mizukoshi, T.; Ueda, T.; Iwai, H.; Shimada, I. Phosphorylation-induced conformation of beta2-adrenoceptor related to arrestin recruitment revealed by NMR. *Nat. Commun.* **2018**, *9*, 194. [CrossRef] [PubMed]

157. Sooklal, C.R.; Lopez-Alonso, J.P.; Papp, N.; Kanelis, V. Phosphorylation Alters the Residual Structure and Interactions of the Regulatory L1 Linker Connecting NBD1 to the Membrane-Bound Domain in SUR2B. *Biochemistry* **2018**, *57*, 6278–6292. [CrossRef] [PubMed]

158. Teriete, P.; Thai, K.; Choi, J.; Marassi, F.M. Effects of PKA phosphorylation on the conformation of the Na,K-ATPase regulatory protein FXYD1. *Biochim. Biophys. Acta* **2009**, *1788*, 2462–2470. [CrossRef]

159. Xiang, S.; Gapsys, V.; Kim, H.Y.; Bessonov, S.; Hsiao, H.H.; Mohlmann, S.; Klaukien, V.; Ficner, R.; Becker, S.; Urlaub, H.; et al. Phosphorylation drives a dynamic switch in serine/arginine-rich proteins. *Structure* **2013**, *21*, 2162–2174. [CrossRef]

160. Selenko, P.; Gregorovic, G.; Sprangers, R.; Stier, G.; Rhani, Z.; Kramer, A.; Sattler, M. Structural basis for the molecular recognition between human splicing factors U2AF65 and SF1/mBBP. *Mol. Cell* **2003**, *11*, 965–976. [CrossRef]

161. Manceau, V.; Swenson, M.; Le Caer, J.P.; Sobel, A.; Kielkopf, C.L.; Maucuer, A. Major phosphorylation of SF1 on adjacent Ser-Pro motifs enhances interaction with U2AF65. *FEBS J.* **2006**, *273*, 577–587. [CrossRef]

162. Wang, W.; Maucuer, A.; Gupta, A.; Manceau, V.; Thickman, K.R.; Bauer, W.J.; Kennedy, S.D.; Wedekind, J.E.; Green, M.R.; Kielkopf, C.L. Structure of phosphorylated SF1 bound to U2AF(6)(5) in an essential splicing factor complex. *Structure* **2013**, *21*, 197–208. [CrossRef]

163. Zhang, Y.; Madl, T.; Bagdiul, I.; Kern, T.; Kang, H.S.; Zou, P.; Mausbacher, N.; Sieber, S.A.; Kramer, A.; Sattler, M. Structure, phosphorylation and U2AF65 binding of the N-terminal domain of splicing factor 1 during 3'-splice site recognition. *Nucleic Acids Res.* **2013**, *41*, 1343–1354. [CrossRef]

164. Hamelberg, D.; Shen, T.; McCammon, J.A. A proposed signaling motif for nuclear import in mRNA processing via the formation of arginine claw. *Proc. Natl. Acad. Sci. USA* **2007**, *104*, 14947–14951. [CrossRef]

165. Kumar, P.; Chimenti, M.S.; Pemble, H.; Schonichen, A.; Thompson, O.; Jacobson, M.P.; Wittmann, T. Multisite phosphorylation disrupts arginine-glutamate salt bridge networks required for binding of cytoplasmic linker-associated protein 2 (CLASP2) to end-binding protein 1 (EB1). *J. Biol. Chem.* **2012**, *287*, 17050–17064. [CrossRef]

166. Thapar, R. Structural basis for regulation of RNA-binding proteins by phosphorylation. *ACS Chem. Biol.* **2015**, *10*, 652–666. [CrossRef]

167. Gogl, G.; Kornev, A.P.; Remenyi, A.; Taylor, S.S. Disordered Protein Kinase Regions in Regulation of Kinase Domain Cores. *Trends Biochem. Sci.* **2019**. [CrossRef]

168. Fletcher, C.M.; McGuire, A.M.; Gingras, A.C.; Li, H.; Matsuo, H.; Sonenberg, N.; Wagner, G. 4E binding proteins inhibit the translation factor eIF4E without folded structure. *Biochemistry* **1998**, *37*, 9–15. [CrossRef]

169. Fletcher, C.M.; Wagner, G. The interaction of eIF4E with 4E-BP1 is an induced fit to a completely disordered protein. *Protein Sci.* **1998**, *7*, 1639–1642. [CrossRef]

170. Lukhele, S.; Bah, A.; Lin, H.; Sonenberg, N.; Forman-Kay, J.D. Interaction of the eukaryotic initiation factor 4E with 4E-BP2 at a dynamic bipartite interface. *Structure* **2013**, *21*, 2186–2196. [CrossRef]

171. Gingras, A.C.; Gygi, S.P.; Raught, B.; Polakiewicz, R.D.; Abraham, R.T.; Hoekstra, M.F.; Aebersold, R.; Sonenberg, N. Regulation of 4E-BP1 phosphorylation: A novel two-step mechanism. *Genes Dev.* **1999**, *13*, 1422–1437. [CrossRef]

172. Bah, A.; Vernon, R.M.; Siddiqui, Z.; Krzeminski, M.; Muhandiram, R.; Zhao, C.; Sonenberg, N.; Kay, L.E.; Forman-Kay, J.D. Folding of an intrinsically disordered protein by phosphorylation as a regulatory switch. *Nature* **2015**, *519*, 106–109. [CrossRef]

173. Tanford, C. How protein chemists learned about the hydrophobic factor. *Protein Sci.* **1997**, *6*, 1358–1366. [CrossRef]

174. Mitrea, D.M.; Kriwacki, R.W. Regulated unfolding of proteins in signaling. *FEBS Lett.* **2013**, *587*, 1081–1088. [CrossRef]

175. Schultz, J.E.; Natarajan, J. Regulated unfolding: A basic principle of intraprotein signaling in modular proteins. *Trends Biochem. Sci.* **2013**, *38*, 538–545. [CrossRef]

176. Kumar, A.; Gopalswamy, M.; Wolf, A.; Brockwell, D.J.; Hatzfeld, M.; Balbach, J. Phosphorylation-induced unfolding regulates p19(INK4d) during the human cell cycle. *Proc. Natl. Acad. Sci. USA* **2018**, *115*, 3344–3349. [CrossRef]

177. Palmer, A.G., 3rd. Probing molecular motion by NMR. *Curr. Opin. Struct. Biol.* **1997**, *7*, 732–737. [CrossRef]

178. Mittermaier, A.K.; Kay, L.E. Observing biological dynamics at atomic resolution using NMR. *Trends Biochem. Sci.* **2009**, *34*, 601–611. [CrossRef]

179. Kovermann, M.; Rogne, P.; Wolf-Watz, M. Protein dynamics and function from solution state NMR spectroscopy. *Q. Rev. Biophys.* **2016**, *49*, e6. [CrossRef]

180. Theillet, F.X.; Binolfi, A.; Frembgen-Kesner, T.; Hingorani, K.; Sarkar, M.; Kyne, C.; Li, C.; Crowley, P.B.; Gierasch, L.; Pielak, G.J.; et al. Physicochemical properties of cells and their effects on intrinsically disordered proteins (IDPs). *Chem. Rev.* **2014**, *114*, 6661–6714. [CrossRef]

181. Tyrrell, J.; Weeks, K.M.; Pielak, G.J. Challenge of mimicking the influences of the cellular environment on RNA structure by PEG-induced macromolecular crowding. *Biochemistry* **2015**, *54*, 6447–6453. [CrossRef]

182. Barbieri, L.; Luchinat, E.; Banci, L. Protein interaction patterns in different cellular environments are revealed by in-cell NMR. *Sci. Rep.* **2015**, *5*, 14456. [CrossRef]

183. Mu, X.; Choi, S.; Lang, L.; Mowray, D.; Dokholyan, N.V.; Danielsson, J.; Oliveberg, M. Physicochemical code for quinary protein interactions in Escherichia coli. *Proc. Natl. Acad. Sci. USA* **2017**, *114*, E4556–E4563. [CrossRef]

184. Smith, A.E.; Zhou, L.Z.; Gorensek, A.H.; Senske, M.; Pielak, G.J. In-cell thermodynamics and a new role for protein surfaces. *Proc. Natl. Acad. Sci. USA* **2016**, *113*, 1725–1730. [CrossRef]

185. McConkey, E.H. Molecular evolution, intracellular organization, and the quinary structure of proteins. *Proc. Natl. Acad. Sci. USA* **1982**, *79*, 3236–3240. [CrossRef]

186. Wirth, A.J.; Gruebele, M. Quinary protein structure and the consequences of crowding in living cells: Leaving the test-tube behind. *Bioessays* **2013**, *35*, 984–993. [CrossRef]

187. Cohen, R.D.; Pielak, G.J. A cell is more than the sum of its (dilute) parts: A brief history of quinary structure. *Protein Sci.* **2017**, *26*, 403–413. [CrossRef]

188. Cohen, R.D.; Guseman, A.J.; Pielak, G.J. Intracellular pH modulates quinary structure. *Protein Sci.* **2015**, *24*, 1748–1755. [CrossRef]

189. Cohen, R.D.; Pielak, G.J. Electrostatic Contributions to Protein Quinary Structure. *J. Am. Chem. Soc.* **2016**, *138*, 13139–13142. [CrossRef]

190. Cohen, R.D.; Pielak, G.J. Quinary interactions with an unfolded state ensemble. *Protein Sci.* **2017**, *26*, 1698–1703. [CrossRef]

191. DeMott, C.M.; Majumder, S.; Burz, D.S.; Reverdatto, S.; Shekhtman, A. Ribosome Mediated Quinary Interactions Modulate In-Cell Protein Activities. *Biochemistry* **2017**, *56*, 4117–4126. [CrossRef]

192. Diniz, A.; Dias, J.S.; Jimenez-Barbero, J.; Marcelo, F.; Cabrita, E.J. Protein-Glycan Quinary Interactions in Crowding Environment Unveiled by NMR Spectroscopy. *Chemistry* **2017**, *23*, 13213–13220. [CrossRef]

193. Kyne, C.; Jordon, K.; Filoti, D.I.; Laue, T.M.; Crowley, P.B. Protein charge determination and implications for interactions in cell extracts. *Protein Sci.* **2017**, *26*, 258–267. [CrossRef]

194. Majumder, S.; Xue, J.; DeMott, C.M.; Reverdatto, S.; Burz, D.S.; Shekhtman, A. Probing protein quinary interactions by in-cell nuclear magnetic resonance spectroscopy. *Biochemistry* **2015**, *54*, 2727–2738. [CrossRef]

195. Monteith, W.B.; Cohen, R.D.; Smith, A.E.; Guzman-Cisneros, E.; Pielak, G.J. Quinary structure modulates protein stability in cells. *Proc. Natl. Acad. Sci. USA* **2015**, *112*, 1739–1742. [CrossRef]

196. Fonin, A.V.; Darling, A.L.; Kuznetsova, I.M.; Turoverov, K.K.; Uversky, V.N. Intrinsically disordered proteins in crowded milieu: When chaos prevails within the cellular gumbo. *Cell Mol. Life Sci.* **2018**, *75*, 3907–3929. [CrossRef]

197. Theillet, F.X.; Binolfi, A.; Bekei, B.; Martorana, A.; Rose, H.M.; Stuiver, M.; Verzini, S.; Lorenz, D.; van Rossum, M.; Goldfarb, D.; et al. Structural disorder of monomeric alpha-synuclein persists in mammalian cells. *Nature* **2016**, *530*, 45–50. [CrossRef]

198. Sarkar, M.; Smith, A.E.; Pielak, G.J. Impact of reconstituted cytosol on protein stability. *Proc. Natl. Acad. Sci. USA* **2013**, *110*, 19342–19347. [CrossRef]

International Journal of
Molecular Sciences

Review

Protein Structure Determination in Living Cells

Teppei Ikeya [1,*], Peter Güntert [1,2,3] and Yutaka Ito [1,*]

[1] Department of Chemistry, Graduate School of Science, Tokyo Metropolitan University, 1-1 Minamiosawa, Hachioji, Tokyo 192-0397, Japan

[2] Institute of Biophysical Chemistry and Center for Biomolecular Magnetic Resonance, Goethe University Frankfurt, 60438 Frankfurt am Main, Germany; guentert@em.uni-frankfurt.de

[3] Laboratory of Physical Chemistry, ETH Zürich, 8093 Zürich, Switzerland; peter.guentert@phys.chem.ethz.ch

* Correspondence: tikeya@tmu.ac.jp (T.I.); ito-yutaka@tmu.ac.jp (Y.I.); Tel.: +81-42-677-2545 (Y.I.)

Received: 26 April 2019; Accepted: 15 May 2019; Published: 17 May 2019

Abstract: To date, in-cell NMR has elucidated various aspects of protein behaviour by associating structures in physiological conditions. Meanwhile, current studies of this method mostly have deduced protein states in cells exclusively based on 'indirect' structural information from peak patterns and chemical shift changes but not 'direct' data explicitly including interatomic distances and angles. To fully understand the functions and physical properties of proteins inside cells, it is indispensable to obtain explicit structural data or determine three-dimensional (3D) structures of proteins in cells. Whilst the short lifetime of cells in a sample tube, low sample concentrations, and massive background signals make it difficult to observe NMR signals from proteins inside cells, several methodological advances help to overcome the problems. Paramagnetic effects have an outstanding potential for in-cell structural analysis. The combination of a limited amount of experimental in-cell data with software for *ab initio* protein structure prediction opens an avenue to visualise 3D protein structures inside cells. Conventional nuclear Overhauser effect spectroscopy (NOESY)-based structure determination is advantageous to elucidate the conformations of side-chain atoms of proteins as well as global structures. In this article, we review current progress for the structure analysis of proteins in living systems and discuss the feasibility of its future works.

Keywords: protein structure determination 1; non-uniform sampling 2; spectrum reconstruction 3; structural calculation 4; paramagnetic effects

---

1. Introduction

More than 15 years have passed since Dötsch and coworkers demonstrated the first NMR spectrum of a small protein (NmerA) in living *Escherichia coli* cells in 2001 [1]. In-cell NMR has extended from prokaryotic to eukaryotic cells and has become the only tool to investigate protein behaviour inside cells at atomic resolution [2–4]. It has an established status as one of the biological applications of solution or solid-state NMR. So far, the method has uncovered various remarkable aspects of protein behaviour in cells or molecular crowding environments [5,6]. In intracellular or molecular crowding environments, several effects, which are generally ignored in diluted solution, such as the excluded-volume effect and nonspecific interactions, dictate protein stability and conformation. It has been proposed that the excluded-volume effect promotes compact forms of proteins [7], and that nonspecific interactions with other molecules inside cells invoke opposite effects [8,9]. Among numerous findings in the complex crowding environments, it is particularly interesting that the living cell environment notably decreases the folding stability of proteins. In-cell NMR H/D exchange experiments of human ubiquitin with three alanine mutations (L8A, I44A and V70A; referred to as ubiquitin 3A) revealed that the exchange rate of backbone amide hydrogens with solvent water was 15–20 times faster in HeLa cells than in diluted solution, demonstrating that the protein fold of ubiquitin 3A was destabilised in HeLa cells [10]. In-cell NMR studies using peculiar proteins, which were in equilibrium between a folded and unfolded conformation in diluted solution, showed that their folded states were destabilised inside mammalian and bacterial cells and the equilibrium was shifted towards the unfolded state in the cells [11,12]. Danielsson et al. [11] studied the thermodynamics of the I35A mutant of SOD1^{barrel} (superoxide dismutase 1) in mammalian (A2780) and bacterial cells, and Smith et al. [12] performed ^{19}F NMR measurements for the 7-kDa globular N-terminal Src homology 3 (SH3) domain of the *Drosophila* signal transduction protein drk (downstream of receptor kinase) in *E. coli* cells. Although these are fairly different samples and experimental conditions, both suggested that destabilised proteins *in vitro* become more unstable in prokaryotic and eukaryotic cells. Moreover, intrinsically disordered proteins (IDPs) in disordered states *in vitro* persisted in disordered conditions even in the crowded cellular environment [13,14]. These results suggest that many destabilised proteins are promoted to more unstable states in intracellular environments.

While in-cell NMR is the principal tool to elucidate a protein's natural behaviour in physiological conditions, the lack of structural information for the proteins in cells makes it difficult to fully understand the mechanisms of protein stability and details of conformational differences to those in diluted solution. In-cell NMR studies based only on chemical shift changes and peak intensities are limited to provide rough descriptions of proteins in cells based on prior knowledge. Thus, the next interest is to obtain 'direct' structural information for understanding how the 3D structures or dynamics of biomolecules are existing in living environments, and how they differ from the state in dilute *in vitro* solution. Despite the remarkable feature of NMR which permits to access conformational information from individual atoms of biomacromolecules and to determine their 3D conformations, 3D protein structure determinations in cells are still very few. This is due to the fact that it is still not straightforward to collect sufficient interatomic distance and angle information from in-cell NMR data because of the low signal-to-noise (S/N) ratio and an enormous number of background signals. The background signals are crucial issues particularly when using the system of intrinsic overexpression of proteins in cells [15,16], or observing ^{13}C resonance spectra so that they contain many noise peaks derived from the natural abundance of ^{13}C sources in cells. Despite these difficulties, several in-cell NMR studies tackled the problems and current progress allows elucidating structural details of proteins inside cells. In this article, we introduce several studies which have yielded direct structural information for biomolecules in living cells, and further determined accurate 3D structures based on measurements of 3D NOESY and triple-resonance NMR spectra. Finally, we discuss the current challenges which must be solved in the next years and feature perspectives for protein structure determination by in-cell NMR.

2. Paramagnetic NMR

Conventional NMR protein structure determinations usually utilise interatomic distance information derived from nuclear Overhauser effects (NOEs) [17]. However, because of abundant background signals, the short lifetime of cells in an NMR sample tube, and low concentrations of the proteins of interest, it is not trivial to record 3D NOESY-type spectra and to collect a sufficient quantity of NOE-derived distance restraints. Whilst recent progress described in the next section can overcome these problems, paramagnetic effects, such as paramagnetic relaxation enhancements (PRE), pseudocontact shifts (PCS), and residual dipolar couplings (RDC), provide alternative or complementary structural data to the NOE-derived distances [18–20]. They are particularly useful for unstable samples, such as living cells in an NMR sample tube, because structural information can be collected even from 2D ^{1}H–^{15}N or ^{1}H–^{13}C correlation spectra, while the NOESY-type experiments require 3D measurements to reduce overlaps among signals of the target protein or with background from cells. It should be noted that the PRE and PCS effects provide long-range structure information from a metal centre for distances up to about 40 Å [18], and RDC provides information on the orientation of bonds for scalar coupled spins relative to the static magnetic field. A drawback for observing the paramagnetic effects is the need to incorporate lanthanide-binding tags (LBTs) to the proteins unless they have a strong natural affinity for paramagnetic lanthanide ions. Either the metal-binding proteins or LBTs must have sufficiently strong affinities to lanthanide ions possessing cytotoxicity. As general issues for the observation of PCS, including *in vitro* experiments, the LBTs also require a stable, covalent linker to the protein, which should be short to limit their flexibility so as to achieve accurate structural information. The tags should maintain only one stereoisomer and minimise the structural distortions of proteins by introducing them. For in-cell NMR experiments, disulphide bonds, which are often used in chemical modifications to proteins, cannot be adopted because of the intracellular reducing environment. The same is true for the conjugating nitroxide radicals with disulphide bonds that are commonly used for obtaining PREs *in vitro*. Recently, several new LBTs have been developed for the acquisition of PRE, PCS, and RDC data inside cells [21–25] (Table 1). An approach to suppress the mobility of LBTs is to employ steric bulk chelators with a relatively short linker. Häussinger et al. proposed a stable LBT yielding very large PCSs (beyond 5 ppm), referred to as DOTA-M8 SPy, which is mainly composed of a DOTA (tetraxetan; 1,4,7,10-tetraazacyclododecane-1,4,7,10-tetraacetic acid) framework and eight methyl groups [26]. The eight methyl groups result in both hydrophilic and hydrophobic faces of the LBT. This allows, in addition to the covalent linker, for a secondary, noncovalent attachment to proteins due to hydrophobic interactions. The low mobility of this LBT due to the steric overcrowding and hydrophobic interactions to proteins yielded large PCSs as well as an extremely high affinity of the DOTA framework to lanthanides with a binding constant of the order of 10^{-25} to 10^{-27} M. Hikone et al. improved DOTA-M8 SPy for obtaining structural information from in-cell NMR by altering the fragile disulfide linkage of the original DOTA-M8 SPy into a carbamidemethyl (CAM) group that is stable in reducing intracellular environments (henceforth referred as DOTA-M8-CAM-I) [24]. DOTA-M8-CAM-I was attached to ubiquitin 3A with two different cysteine mutations (K6C and S57C). Incorporating the $[Dy^{3+}(DOTA-M8-CAM-I)]$-tagged ubiquitin mutants into HeLa cells, this permitted observation of relatively large PCSs in order to obtain structural information for a protein in human cultured cells. In the meantime, Müntener et al. investigated the reactivity and stability of several LBTs possessing a DOTA-M7Py framework with a linker comprising a thioether bond that is irreversible under reducing conditions including intracellular environments [22,25]. Among them, M7PyThiazole-SO$_2$Me-DOTA showed high stability, reactivity, and strong PCSs and RDCs. It efficiently reacted to more than 99% within 5 min at pH 7.0 and 295 K with a small peptide (Leu-Cys-Asp), which was the identical sequence as the tagging site of ubiquitin 3A S57C. It also tagged ubiquitin 3A S57C to an extent >95% and did not hydrolyse at pH 7.0 and room temperature. It yielded large PCSs and RDCs (up to 10 ppm and 32 Hz) with Dy^{2+} when conjugated with ubiquitin 3A S57C and K48C, and human carbonic anhydrase II (hCA-II) S166C/C206S. This high performance was attributed to the rigidity of the linker consisting of a sulphide bond between the pyridine thiazole ring and a cysteine residue that is probably due to steric clashes of the tag with the protein. Although in-cell NMR using this LBT has not been reported yet, these new stable tags for reducing environments inside cells are good candidates for obtaining accurate structural information in future in-cell NMR studies.

While in principle PCS is one of the only few sources of reliable experimental data containing direct structural information of proteins for in-cell NMR, it is still difficult to determine *de novo* protein structures exclusively from PCS data. Comprehensive reviews of the theory and applications of PCS *in vitro* can be found in articles by Bertini, Otting, and Meiler et al. [19,30,31]. The PCS $\Delta\delta^{PCS}$ is related to the structure and the magnetic susceptibility tensor by

$$\Delta\delta^{PCS} = \frac{1}{12\pi r^3}\left[\Delta\chi_{ax}\left(3\cos^2\theta - 1\right) + \frac{3}{2}\Delta\chi_{rh}\sin^2\theta\cos 2\phi\right] \tag{1}$$

where r is the distance between the lanthanide ion and the atom for which the PCS is observed, θ and ϕ are polar coordinates of the atom with respect to the magnetic susceptibility tensor $\Delta\chi$ of the lanthanide, and $\Delta\chi_{ax}$ and $\Delta\chi_{rh}$ are the axial and rhombic components of $\Delta\chi$, respectively. It contains not only distances between a metal centre and individual atoms, but also a priori unknown magnetic susceptibility tensor parameters, as well as ambiguity due to the flexibility of the LBTs and experimental errors [32,33]. The determination of the tensor components without prior 3D structure information is not straightforward compared to the data analysis for NOEs and PREs. Also, large chemical shift differences between diamagnetic and paramagnetic data can make it difficult to achieve resonance assignments. The signals weakened by the paramagnetic relaxation enhancement effect and massive background signals hinder the analysis of these data as well. Hence, the shifted signals are generally assigned by substituting several lanthanide ions with different magnetic susceptibilities so as to gradually alter the magnitude of the shifts and depicting their trajectories in the spectra. Moreover, it is necessary to collect PCS data from several different paramagnetic centres, or mutation sites, in order to obtain sufficient structural information for most of the protein. Considering that in-cell NMR experiments require a substantial amount of protein samples and cells, it is demanding to perform sufficient PCS experiments for the resonance assignment and data collection required for a structure determination.

Table 1. Lanthanide-binding tags (LBTs) proposed for inducing paramagnetic effects in living cells.

Name	Chemical Structure [1]	Reported Paramagnetic Effects	Linker	Reference	Commercially Available [CAS] [2]
DOTA-M8-CAM-I		PCS	*N*-propylene-acetamide	[24]	no
4PhSO$_2$-PyMTA		PCS/PRE	pyridine	[23,27]	yes
DOTA-maleimide		PRE	*N*-ethylene-maleimide	[28]	yes [1006711-09-5]
DOTA-M7Py		PCS	pyridine	[22]	no
DO2A		solvent PRE	—	[29]	yes [112193-75-6]
M7PyThiazole-SO2Me-DOTA		PCS/RDC	pyridine thiazole	[25]	no

[1] M denotes lanthanoid ions, [2] Chemical Abstracts Service (CAS) registry number, if available.

Thus, two distinct groups, Müntener et al. [22] and Pan et al. [23], employed an empirical approach for protein structure determination using PCS- or GPS-ROSETTA, which relies on the ROSETTA software that was originally developed for *ab initio* protein structure prediction from amino acid sequences [34,35]. At nearly the same time, the two groups performed structure determinations of *Streptococcus* protein G B1 domain (57 a.a., 7 kDa; henceforth referred to as GB1) by in-cell NMR using *Xenopus laevis* oocytes.

Müntener et al. [22] utilized DOTA-M7Py which maintains several remarkable features: (1) strong affinity ($K_d < 10^{-25}$ M) toward lanthanides due to the DOTA framework, (2) exclusively the square antiprismatic $\Lambda(\delta\delta\delta\delta)$ stereo-configuration for the 4S,3R-Lu derivative, (3) a short linker with a nonreducible thioether bond, and (4) both hydrophilic and hydrophobic properties for reducing mobility in a similar manner as DOTA-M8 SPy. The measurements of in-cell PCSs and RDCs using this tag were achieved by microinjection into *Xenopus* oocytes for an intracellular concentration of GB1 of about 50 μM. The PCSs of ^{1}H and ^{15}N, and ^{1}H–^{15}N RDCs were respectively collected for Tm^{3+} and Tb^{3+} at three Cys-mutation sites. Because ROSETTA has the ability to predict 3D protein structures from sequences, the authors tested the prediction performance of GPS-ROSETTA under three input conditions: no experimental data, only PCS, and PCS and RDC. Their results demonstrated that GPS-ROSETTA using the PCS and RDC data provided accurate global structures of GB1 in *Xenopus* oocytes with a C^{α} RMSD of 0.64 Å to the crystal structure (Protein Data Bank; PDB code: 2QMT), which was not possible by structure prediction with ROSETTA from the sequence alone. They concluded that this approach was sufficiently accurate to determine well-defined protein structures, and the overall structural features of GB1 in Xenopus oocytes were similar to those observed *in vitro*.

Pan et al. [23] performed the 3D structure determination with PCS in *Xenopus* oocytes using GPS-ROSETTA, by a similar approach as described above except that the LBT 4PhSO$_2$-PyMTA [27] was used. Features of the commercially available 4PhSO$_2$-PyMTA are that it has a short and stable thioether bond in the linker to the protein, and a simple chemical structure compared to the compounds with the DOTA-framework. Using PCSs at two Cys-mutation sites with Tb^{3+}, Tm^{3+}, and Yb^{3+}, they performed the modelling of the GB1 structures in cells by GPS-ROSETTA. The structure by ROSETTA with the PCS data has a C^{α} RMSD of 1.0 Å from the crystal structure (PDB code: 2QMT) and the 25 lowest energy structures were a well-converged with less than a C^{α} RMSD of 0.15 Å from the lowest energy one. The method presented sufficiently accurate structures of GB1 which remained unchanged in the cellular environment despite the notable structural variations for residues 8–12 in a loop. Hence, these results demonstrated by the two groups suggest that the combination approach of in-cell data with software for *ab initio* protein structure prediction, such as ROSEETA, would be a powerful tool to visualise 3D protein structures inside cells.

Compared to the difficulty for the collection and analysis of PCS data of in-cell NMR, the acquisition of structural information from PRE may be advantageous from some perspectives [19,36]: (1) several chemically stable tags under reducing environment are commercially available, (2) the calibration of interatomic distance from PRE data is simpler due to its isotropic effect, and (3) PREs are relatively tolerant against tag flexibility compared to PCS [19,36]. In the case of a hydrogen atom, the intensity ratio of a particular proton in paramagnetic/diamagnetic spectra, $I_{\text{para}}/I_{\text{dia}}$, is approximated from the transverse relaxation rates of a diamagnetic and paramagnetic spin, R_2 and R_2^{sp}, respectively [37]:

$$\frac{I_{\text{para}}}{I_{\text{dia}}} = \frac{R_2 \, \exp\left(-R_2^{\text{sp}} t\right)}{R_2 + R_2^{\text{sp}}} \tag{2}$$

where t is the total INEPT (insensitive nuclei enhanced by polarization transfer) evolution time in the case of a heteronuclear single quantum coherence (HSQC) spectrum. The paramagnetically enhanced transverse relaxation rate R_2^{sp} is converted into a distance from the metal centre by use of the following equation:

$$R_2^{\text{sp}} = \frac{K}{r^6}\left(4\tau_c + \frac{4\tau_c}{1 + \omega_h^2 \tau_c^2}\right) \tag{3}$$

where the constant K is 1.23×10^{-32} cm^6 s^{-2}, r is the distance between a lanthanide ion and a hydrogen atom, τ_c is the correlation time for the lanthanide, and ω_h is the Larmor frequency of the proton. Although R_2 and τ_c have to be estimated from additional relaxation experiments to be precise, in-cell NMR studies usually employ approximated values obtained from the line width at half-height of peaks for simplicity. This approximation makes it easy to convert PRE data into distance restraints albeit it sacrifices some of their accuracy. It is also advantageous that some tags for PRE measurements in cells are commercially available. Theillet et al. [28] attached a DOTA-maleimide tag with a Gd^{3+} ion to a single cysteine of a representative IDP, α-synuclein, and observed PREs in human cultured cells [28]. The incorporation of LBTs into proteins via maleimide coupling is chemically stable in reducing environments. In the article, the authors showed that the α-synuclein conformation in cells is similarly compact as *in vitro* and that the character of intrinsic disorder is sustainable inside cells.

3. *De Novo* in-Cell Protein Structure Determination

3.1. De novo Structure Determination in Prokaryotic Cells

As described above, in-cell NMR has made it possible to obtain structural information in order to infer partial conformations of proteins in cells with prior knowledge. Meanwhile, it is also indispensable for the analysis of protein behaviour in cells to accomplish complete structure determinations including side-chain atoms. For instance, side-chain conformations are thought to be predominantly affected by the intracellular environment, which allows to understand the functions of proteins and extend the method to other applications such as structure-based drug discovery. It remains necessary to achieve *de novo* 3D protein structure determination in living cells using NOE-derived distance restraints between side-chains. However, as already mentioned, there are several obstacles to collect a sufficient number of distance restraints from NOESY-type spectra in intracellular environments, e.g., the low concentration of target molecules in cells, the short lifetime of cells, severe background signals from other components of the cells, etc. To date, NMR observations of isotopically labelled-proteins inside cells have been achieved principally by two approaches: intrinsic overexpression of proteins in cells [1,16,38,39] and incorporation of stable isotope-labelled molecules by importing them through the cellular membrane [10,28,40]. The system of intrinsic protein overexpression has advantages in terms of the protein concentration in cells and practical experimental simplicity. For instance, the concentration of the *T. thermophilus* HB8 TTHA1718 gene product (66 a.a., 7 kDa) could reach up to 3–4 mM in *E. coli* cells [15]. It also allows regulation of the protein concentration by altering the delay time after induction and the incubation temperature to some extent [41]. In addition, the method can easily repeat in-cell NMR measurements owing to the omission of protein purification steps. Although the background signals derived from cells are a critical issue in this approach, the desirable features permitted to measure ^{13}C-separated, ^{15}N-separated NOESYs, and 3D triple-resonance spectra for backbone and side-chain resonance assignments. Indeed, protein structure determinations by in-cell NMR currently reported have been achieved exclusively by the system of intrinsic protein overexpression. The first *de novo* protein structure determination by in-cell NMR utilised an overexpression system in *E. coli* [41]. The short lifetime of cells in an NMR sample tube was a challenge for the measurement of NOESY spectra, because *E. coli* cells died or started to release target proteins within approximately 6 h, while it generally takes a couple of days to record 3D NMR spectra with high S/N ratio and resolution. Among the many approaches to address this issue, the most robust and straightforward method is to utilise nonuniform sampling (NUS) in combination with spectral reconstruction by non-Fourier transform methods [42]. The sparse sampling by NUS allows to cover the entire original experimental data matrix ensuring sufficient peak intensity within the cell lifetime. In the first in-cell structure determination, the NUS data was reconstructed by interpolating the missing points with a maximum entropy (MaxEnt) approach [43,44]. This enabled the assignment of the backbone and a majority of the side-chain atoms, as well as collecting a sufficient number of NOE distance restraints for the protein TTHA1718. The resulting structure was well-defined with a backbone root-mean-square deviation (RMSD) below 1.0 Å and is similar to the *in vitro* structure with a backbone RMSD between the two structures of 1.2 Å. Slight structural differences were observed in the putative heavy metal binding loop where chemical shift differences between in-cell and *in vitro* reflected possible metal binding. The authors discussed that the interactions with metal ions in the *E. coli* cytosol or the effects of viscosity and intracellular molecular crowding might affect the conformation of this region.

Later on, the same group improved the procedure for the *de novo* in-cell protein structure determination with three methodological advances composed of improved NMR data processing of NUS data, automated chemical shift assignment, and robust structure calculation with Bayesian inference [45]. The new procedure permitted 3D protein structure determinations with much lower intracellular protein concentrations and even without indirect restraints such as hydrogen bond information. The structure of the protein GB1 in living *E. coli* cells was determined at an order of magnitude lower concentration (approximately 250 μM) in the NMR tubes than in the original report for

TTHA1718 (3–4 mM). This is comparable to a physiologically natural environment, where the maximal natural concentration of a protein in normal cells is a few dozen to hundreds of µM [46,47]. The NMR data processing for the indirect dimensions of 3D NMR spectra employed the Quantitative Maximum Entropy (QME) method [16] instead of the conventional 2D MaxEnt implemented in the program Azara [48] that had been used for the previous structure determination. Chemical shifts of in-cell GB1 were assigned by combining conventional manual analysis with an automated assignment procedure using the FLYA (fully automated assignment) algorithm [49]. A remarkable feature of FLYA is that it enables resonance assignments exclusively from NOESY-type NMR spectra [50–52]. Using NOESY spectra was crucial for obtaining side-chain assignments because faster transverse relaxation of in-cell samples hinders the collection of a sufficient number of signals from through-bond spectra, e.g., H(CCCO)NH, for side-chain resonance assignments. NOESY spectra, on the other hand, contained a considerable number of signals from the side-chains. Structure calculations were performed employing the program CYANA (combined assignment and dynamics algorithm for NMR applications) with the CYBAY (CYANA Bayesian inference) module [53], which was able to extract a maximum of structural information from the limited and ambiguous experimental NOESY data collected in living cells. The GB1 structure ensemble of 1900 conformers calculated by CYBAY is well defined with an average backbone RMSD of 0.43 Å to the mean coordinates (Figure 1). The backbone RMSD between its mean structure and the *in vitro* structure was 1.18 Å. A loop and the end of a β-strand (residues 11–14) showed low RMSDs to the *in vitro* structure for the C^α atoms but higher RMSDs of more than 2.0 Å for the side-chains. These residues coincided well with a region of slightly higher chemical shift differences between the in-cell and *in vitro* samples (residues 10–13). The authors discussed that the structural changes of the side-chains might be due to molecular crowding effects or the intracellular environment, in which the interactions with other negatively charged molecules might result in the structural changes of side-chains. The improved method yielded in-cell TTHA1718 structures that were much better defined than before owing to the additional distance restraints identified by the FLYA analysis of the quantitative maximum entropy (QME)-processed NOESY spectra and the improved distance accuracy by Bayesian inference (Figure 1). The study also confirmed that structural differences were located in three dynamics loop regions (residue 9–12, 26–29 and 44–50) of TTHA1718, which may be affected by the viscosity and macromolecular crowding in the cytosol.

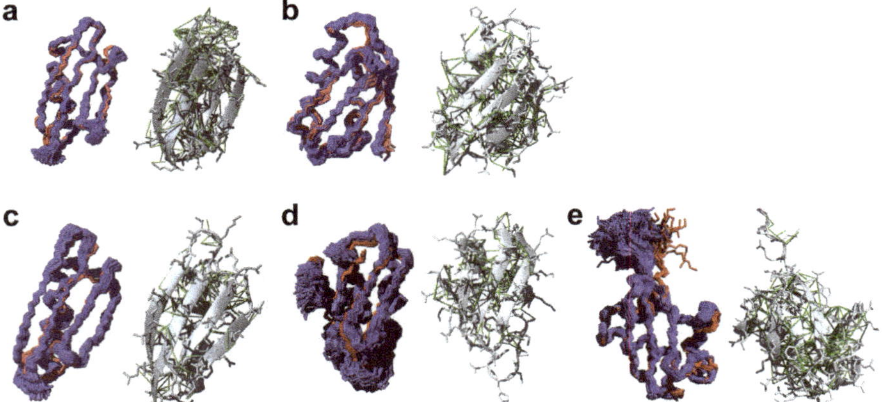

Figure 1. Protein structures determined by in-cell NMR. GB1 (**a**; PDB code: 2N9L) and TTHA1718 (**b**) structures in *E. coli* cells. GB1 (**c**; 5Z4B), TTHA1718 (**d**; 6K1V), and ubiquitin 3A mutant (**e**; 6K1U) structures in sf9 cells. For all structures, the left panels show the backbones of the structure ensemble in cells (blue) and in diluted solution (red). The right panels show the best conformer in cells with side-chains (grey) and NOE distance restraints (green).

3.2. De novo Structure Determination in Eukaryotic Cells

Until very recently, in-cell *de novo* structure determinations were performed exclusively in *E. coli* cells, presumably because the achievable target protein concentration in eukaryotic cells was too low to obtain a sufficient number of NOE-derived distance restraints. In 2019, Tanaka et al. showed the first *de novo* protein structure determinations in living eukaryotic cells using a sf9 cell/baculovirus system, which are based exclusively on information from 3D heteronuclear multidimensional NMR spectra [54]. As model systems, three small- and two medium-sized proteins were chosen: GB1, TTHA1718, ubiquitin 3A, rat calmodulin (148 a.a., 17 kDa), and C-terminally truncated human HRas (residues 1–171, 19 kDa). The concentration of GB1 protein in the sf9 samples was estimated as $50 \pm 12\ \mu M$, which is nearly the maximum protein concentration in intracellular environments, but too low to determine 3D structures by the previous methods. Thus, the authors applied the bioreactor system supplying fresh medium into the NMR tube continuously during the measurements [55–57] to prolong the lifetime of the cells, and hence the NMR measurement time, to at least 24 h with over 90% cell viability. The bioreactor system also removed extracellular proteins released from the cells, ensuring that the in-cell NMR spectra were obtained only from protein inside sf9 cells. All 3D NMR data were sparsely measured by a sampling scheme with a sinusoidal-weighted Poisson distribution, so-called Poisson-gap sampling [58], in order to minimally suppress missing information from large gaps between data points and biased distributions. Subsequently, data were reconstructed by QME. The measurement and processing scheme with the bioreactor, sparse sampling, and QME reconstruction improved significantly the sensitivity of the in-cell spectra. The quality of these spectra allowed to collect NOE-derived distance restraints sufficient for determining high-resolution 3D structures.

For GB1 in sf9 cells, 3D triple-resonance NMR spectra could be measured with high quality for backbone and side-chain resonance assignments, and unambiguous assignments were achieved for approximately 98% of the backbone resonances of GB1 in sf9 cells. The Bayesian inference-assisted structure refinement well defined ensemble structures of GB1 with an average backbone RMSD of 0.51 Å to the mean coordinates (Figure 1). The backbone RMSD between the mean structures in sf9 cells and in diluted solution was 1.61 Å, except for a region composed of a loop and an α-helix (residues 22–26, 28) that shows higher RMSD values around 1.5 Å. These residues coincided well with a region exhibiting chemical shift differences between the in-cell and diluted solution samples (residues 20–24, 27). The relative position of the α-helix in-sf9 structures was tilted away from the β-sheet. The authors concluded that the changes in chemical shift and 3D structure for this region, which interacts through a hydrophobic patch on the protein surface with other molecules nonspecifically, were presumably due to the effects caused by the intracellular environment. A structural difference compared to the structure in diluted solution was also observed in a similar region of the GB1 structure in *E. coli* cells (residues 20–24) [45], and a molecular dynamics study simulating crowded environments [59] suggested that the intracellular environment perturbed the conformation of the region similarly in *E. coli* and sf9 cells. It is interesting to note that the method achieved the observation of minor structural differences even from in-cell NMR spectra with low S/N ratio. However, GB1 is currently the only protein whose structure has been very accurately determined in living eukaryotic cells. Verifications of this finding for other proteins may be needed. In order to elucidate structural differences more quantitatively, it is also essential to validate the in-cell structures in various ways, such as by measuring paramagnetic effects and observing structural dynamics in cells with NMR relaxation experiments.

Structure determinations of Ub3A and TTHA1718 were also performed with distance restraints obtained from 3D NOESY spectra in sf9 cells. The chemical shift assignments for these proteins were transferred from the data in diluted solution based on the knowledge that chemical shift differences for these proteins were small between sf9 cells and diluted solution. The resulting structure ensemble of Ub3A was well-defined with an average backbone RMSD of 0.39 Å to the mean coordinates, and 1.31 Å to that in diluted solution (Figure 1). The structure ensemble of TTHA1718 presented an average backbone RMSD of 0.88 and 2.60 Å to the mean coordinates and to that in diluted solution, respectively. The relatively large RMSDs of TTHA1718 were attributed to the putative metal-binding loop region of residues 9–18, for which only few NOE distance restraints were collected, presumably due to exchange processes related to the binding of various metal ions. Excluding this region, the backbone RMSD to the structure in diluted solution was 1.27 Å.

For calmodulin and HRas in sf9, samples were prepared with methyl- and aromatic-selective ^{1}H/^{13}C- labelling to reduce signal overlap. In both cases, well-resolved 3D ^{13}C-separated NOESY spectra were acquired, indicating that this method is effective for obtaining NOE-derived structural information of proteins with molecular weight over 15 kDa in eukaryotic cells. Comparison of in-cell NMR spectra of calmodulin with those in diluted solution suggested that calmodulin in sf9 cells maintains a conformational state similar to the Mg^{2+}-bound form in diluted solution. This data indicates that the bioreactor system successfully kept sf9 cells healthy and suppressed the stress-induced increase of the intracellular Ca^{2+} concentration, which has been reported in a previous in-cell NMR study using HeLa cells [60]. Spectra of HRas expressed in sf9 cells were similar to those in the 'inactive' guanosine diphosphate (GDP)-bound state, suggesting that the C-terminally truncated human HRas cannot be activated to the guanosine triphosphate (GTP)-bound form at the cell membrane, while GTP molecules bound to HRas during protein synthesis in sf9 will be hydrolysed by its intrinsic GTPase activity during the incubation period. Finally, the authors concluded that backbone and side-chain resonance assignments and 3D structures of proteins of less than 10 kDa size can be determined exclusively from NOE-derived distance restraints acquired in living sf9 cells. For medium-sized proteins, such as calmodulin and HRas, it is possible to obtain high-resolution structural information from in-cell NOESY experiments in combination with selective ^{1}H/^{13}C-labelling.

4. Conclusions and Outlook

In this article, we focused on in-cell NMR studies that yielded direct structural information or determined protein 3D structures in living systems. The experimental and computational techniques outlined in this article provide insight into the 3D structural information for a variety of biological functions in living systems. The paramagnetic effects, such as PRE, PCS, and RDC, provide highly promising data, while NOE-based structure determination is crucial to delineate detailed protein conformations including side-chain atoms. Despite the high interest in studying protein conformations with atomic resolution in cells, to the best of our knowledge, these studies are still very few and limited to a small number of proteins. It might imply that these are delicate methods that are too demanding and laborious for routine applications to many proteins and associated functional studies. However, incorporating the current progress described here, we expect 3D structure analysis by in-cell NMR to play a significant future role in elucidating functionally relevant structure and dynamics at atomic resolution in living cells. Moreover, stereospecific isotope labelling [61,62] and segmental labelling techniques [63–67] would be valuable for future applications of in-cell NMR. State-of-the-art computational methods in automatic resonance assignment using a limited number of spectra [51,52], protein 3D structure prediction [68], sparse modelling such as compressive sensing [69–72], machine learning [73], and molecular dynamics simulation [74] will also contribute to the advancement in this field. Over time, these methods will jointly help to elucidate the functions and behaviour of biomolecules in living systems and find potential application in structure-based drug screening and clinical therapy.

Author Contributions: Conceptualization, writing—original draft preparation, T.I.; writing—review and editing, P.G. and Y.I.

Funding: We gratefully acknowledge financial supports by Scientific Research on Innovative Areas (JP26102538, JP25120003, JP16H00779 to T.I., and JP15H01645, JP16H00847 to Y.I.) and Grants-in-Aid for Scientific Research (JP15K06979 to T.I., JP17K07312 to P.G.) from the Japan Society for the Promotion of Science (JSPS) and by the Funding Program for Core Research for Evolutional Science and Technology (CREST JPMJCR13M3) from Japan Science and Technology Agency (JST).

Acknowledgments: The authors thank Hiromasa Yagi (RIKEN) and Tsutomu Terauchi (Taiyo Nippon Sanso corp.) for useful discussion of paramagnetic effects, and Hirabayashi (Tokyo Metropolitan University) for carefully confirming chemical compounds.

Conflicts of Interest: The authors declare no conflict of interest. The funders had no role in the design of the study; in the collection, analyses, or interpretation of data; in the writing of the manuscript, or in the decision to publish the results.

Abbreviations

HSQC   Heteronuclear Single Quantum Coherence
H/D   Hydrogen/Deuterium
NUS   Non-uniform Sampling
SOD   Superoxide Dismutase
PCS   Pseudo-Contact Shift
RDC   Residual Dipolar Coupling
IDP   Intrinsically Disordered Protein
LBT   Lanthanide-Binding Tag
NUS   Non-Uniform Sampling
NOE   Nuclear Overhauser Effect
GB1   protein G B1 domain
PRE   Paramagnetic Relaxation Enhancement
PDB   Protein Data Bank

References

1. Serber, Z.; Keatinge-Clay, A.T.; Ledwidge, R.; Kelly, A.E.; Miller, S.M.; Dötsch, V. High-resolution macromolecular NMR spectroscopy inside living cells. *J. Am. Chem. Soc.* **2001**, *123*, 2446–2447. [CrossRef]
2. Freedberg, D.I.; Selenko, P. Live cell NMR. *Annu. Rev. Biophys.* **2014**, *43*, 171–192. [CrossRef]
3. Luchinat, E.; Banci, L. In-cell NMR: A topical review. *IUCrJ* **2017**, *4*, 108–118. [CrossRef] [PubMed]
4. Ikeya, T.; Ban, D.; Lee, D.; Ito, Y.; Kato, K.; Griesinger, C. Solution NMR views of dynamical ordering of biomacromolecules. *Biochim. Biophys. Acta Gen. Subj.* **2018**, *1862*, 287–306. [CrossRef] [PubMed]
5. Ellis, R.J. Macromolecular crowding: Obvious but underappreciated. *Trends Biochem. Sci.* **2001**, *26*, 597–604. [CrossRef]
6. Majumder, S.; Xue, J.; DeMott, C.M.; Reverdatto, S.; Burz, D.S.; Shekhtman, A. Probing protein quinary interactions by in-cell nuclear magnetic resonance spectroscopy. *Biochemistry* **2015**, *54*, 2727–2738. [CrossRef] [PubMed]
7. Zhou, H.-X.; Rivas, G.; Minton, A.P. Macromolecular Crowding and Confinement: Biochemical, Biophysical, and Potential Physiological Consequences. *Annu. Rev. Biophys.* **2008**, *37*, 375–397. [CrossRef] [PubMed]
8. Miklos, A.C.; Li, C.G.; Sharaf, N.G.; Pielak, G.J. Volume Exclusion and Soft Interaction Effects on Protein Stability under Crowded Conditions. *Biochemistry* **2010**, *49*, 6984–6991. [CrossRef] [PubMed]
9. Wang, Y.Q.; Li, C.G.; Pielak, G.J. Effects of Proteins on Protein Diffusion. *J. Am. Chem. Soc.* **2010**, *132*, 9392–9397. [CrossRef]
10. Inomata, K.; Ohno, A.; Tochio, H.; Isogai, S.; Tenno, T.; Nakase, I.; Takeuchi, T.; Futaki, S.; Ito, Y.; Hiroaki, H.; et al. High-resolution multi-dimensional NMR spectroscopy of proteins in human cells. *Nature* **2009**, *458*, 106–109. [CrossRef]
11. Danielsson, J.; Mu, X.; Lang, L.; Wang, H.; Binolfi, A.; Theillet, F.X.; Bekei, B.; Logan, D.T.; Selenko, P.; Wennerstrom, H.; et al. Thermodynamics of protein destabilization in live cells. *Proc. Natl. Acad. Sci. USA* **2015**, *112*, 12402–12407. [CrossRef]
12. Smith, A.E.; Zhou, L.Z.; Gorensek, A.H.; Senske, M.; Pielak, G.J. In-cell thermodynamics and a new role for protein surfaces. *Proc. Natl. Acad. Sci. USA* **2016**, *113*, 1725–1730. [CrossRef] [PubMed]
13. Barnes, C.O.; Monteith, W.B.; Pielak, G.J. Internal and global protein motion assessed with a fusion construct and in-cell NMR spectroscopy. *ChemBioChem* **2011**, *12*, 390–391. [CrossRef] [PubMed]
14. Smith, A.E.; Zhou, L.Z.; Pielak, G.J. Hydrogen exchange of disordered proteins in Escherichia coli. *Protein Sci.* **2015**, *24*, 706–713. [CrossRef]
15. Ikeya, T.; Sasaki, A.; Sakakibara, D.; Shigemitsu, Y.; Hamatsu, J.; Hanashima, T.; Mishima, M.; Yoshimasu, M.; Hayashi, N.; Mikawa, T.; et al. NMR protein structure determination in living E. coli cells using nonlinear sampling. *Nat. Protoc.* **2010**, *5*, 1051–1060. [CrossRef]

16. Hamatsu, J.; O'Donovan, D.; Tanaka, T.; Shirai, T.; Hourai, Y.; Mikawa, T.; Ikeya, T.; Mishima, M.; Boucher, W.; Smith, B.O.; et al. High-resolution heteronuclear multidimensional NMR of proteins in living insect cells using a baculovirus protein expression system. *J. Am. Chem. Soc.* **2013**, *135*, 1688–1691. [CrossRef] [PubMed]

17. Wüthrich, K. *NMR of Proteins and Nucleic Acids*; Wiley-Interscience: New York, NY, USA, 1986.

18. Clore, G.M.; Iwahara, J. Theory, practice, and applications of paramagnetic relaxation enhancement for the characterization of transient low-population states of biological macromolecules and their complexes. *Chem. Rev.* **2009**, *109*, 4108–4139. [CrossRef] [PubMed]

19. Otting, G. Protein NMR using paramagnetic ions. *Annu. Rev. Biophys.* **2010**, *39*, 387–405. [CrossRef]

20. Saio, T.; Inagaki, F. *Experimental Approaches of NMR Spectroscopy: Structural Study of Proteins by Paramagnetic Lanthanide Probe Methods*; Springer: New York, NY, USA, 2018; pp. 227–252.

21. Ye, Y.; Liu, X.; Xu, G.; Liu, M.; Li, C. Direct observation of Ca^{2+}-induced calmodulin conformational transitions in intact Xenopus laevis oocytes by ^{19}F NMR spectroscopy. *Angew. Chem. Int. Ed. Engl.* **2015**, *54*, 5328–5330. [CrossRef]

22. Müntener, T.; Häussinger, D.; Selenko, P.; Theillet, F.X. In-Cell Protein Structures from 2D NMR Experiments. *J. Phys. Chem. Lett.* **2016**, *7*, 2821–2825. [CrossRef] [PubMed]

23. Pan, B.B.; Yang, F.; Ye, Y.; Wu, Q.; Li, C.; Huber, T.; Su, X.C. 3D structure determination of a protein in living cells using paramagnetic NMR spectroscopy. *Chem. Commun. (Camb.)* **2016**, *52*, 10237–10240. [CrossRef]

24. Hikone, Y.; Hirai, G.; Mishima, M.; Inomata, K.; Ikeya, T.; Arai, S.; Shirakawa, M.; Sodeoka, M.; Ito, Y. A new carbamidemethyl-linked lanthanoid chelating tag for PCS NMR spectroscopy of proteins in living HeLa cells. *J. Biomol. NMR* **2016**, *66*, 99–110. [CrossRef] [PubMed]

25. Müntener, T.; Kottelat, J.; Huber, A.; Häussinger, D. New Lanthanide Chelating Tags for PCS NMR Spectroscopy with Reduction Stable, Rigid Linkers for Fast and Irreversible Conjugation to Proteins. *Bioconjug. Chem.* **2018**, *29*, 3344–3351. [CrossRef] [PubMed]

26. Häussinger, D.; Huang, J.R.; Grzesiek, S. DOTA-M8: An extremely rigid, high-affinity lanthanide chelating tag for PCS NMR spectroscopy. *J. Am. Chem. Soc.* **2009**, *131*, 14761–14767. [CrossRef] [PubMed]

27. Yang, Y.; Wang, J.T.; Pei, Y.Y.; Su, X.C. Site-specific tagging proteins via a rigid, stable and short thiolether tether for paramagnetic spectroscopic analysis. *Chem. Commun. (Camb.)* **2015**, *51*, 2824–2827. [CrossRef] [PubMed]

28. Theillet, F.X.; Binolfi, A.; Bekei, B.; Martorana, A.; Rose, H.M.; Stuiver, M.; Verzini, S.; Lorenz, D.; van Rossum, M.; Goldfarb, D.; et al. Structural disorder of monomeric alpha-synuclein persists in mammalian cells. *Nature* **2016**, *530*, 45–50. [CrossRef] [PubMed]

29. Theillet, F.X.; Binolfi, A.; Liokatis, S.; Verzini, S.; Selenko, P. Paramagnetic relaxation enhancement to improve sensitivity of fast NMR methods: Application to intrinsically disordered proteins. *J. Biomol. NMR* **2011**, *51*, 487–495. [CrossRef]

30. Bertini, I.; Luchinat, C.; Parigi, G. Magnetic susceptibility in paramagnetic NMR. *Prog. Nucl. Magn. Reson. Spectrosc.* **2002**, *40*, 249–273. [CrossRef]

31. Koehler, J.; Meiler, J. Expanding the utility of NMR restraints with paramagnetic compounds: Background and practical aspects. *Prog. Nucl. Magn. Reson. Spectrosc.* **2011**, *59*, 360–389. [CrossRef]

32. Schmitz, C.; Stanton-Cook, M.J.; Su, X.C.; Otting, G.; Huber, T. Numbat: An interactive software tool for fitting Deltachi-tensors to molecular coordinates using pseudocontact shifts. *J. Biomol. NMR* **2008**, *41*, 179–189. [CrossRef] [PubMed]

33. Suturina, E.A.; Kuprov, I. Pseudocontact shifts from mobile spin labels. *Phys. Chem. Chem. Phys.* **2016**, *18*, 26412–26422. [CrossRef] [PubMed]

34. Bradley, P.; Malmstrom, L.; Qian, B.; Schonbrun, J.; Chivian, D.; Kim, D.E.; Meiler, J.; Misura, K.M.; Baker, D. Free modeling with Rosetta in CASP6. *Proteins* **2005**, *61*, 128–134. [CrossRef] [PubMed]

35. Ovchinnikov, S.; Park, H.; Kim, D.E.; DiMaio, F.; Baker, D. Protein structure prediction using Rosetta in CASP12. *Proteins* **2018**, *86*, 113–121. [CrossRef] [PubMed]

36. Clore, G.M. Practical Aspects of Paramagnetic Relaxation Enhancement in Biological Macromolecules. *Methods Enzymol.* **2015**, *564*, 485–497.

37. Battiste, J.L.; Wagner, G. Utilization of Site-Directed Spin Labeling and High-Resolution Heteronuclear Nuclear Magnetic Resonance for Global Fold Determination of Large Proteins with Limited Nuclear Overhauser Effect Data. *Biochemistry* **2000**, *39*, 5355–5365. [CrossRef]

38. Bertrand, K.; Reverdatto, S.; Burz, D.S.; Zitomer, R.; Shekhtman, A. Structure of proteins in eukaryotic compartments. *J. Am. Chem. Soc.* **2012**, *134*, 12798–12806. [CrossRef]

39. Banci, L.; Barbieri, L.; Bertini, I.; Luchinat, E.; Secci, E.; Zhao, Y.; Aricescu, A.R. Atomic-resolution monitoring of protein maturation in live human cells by NMR. *Nat. Chem. Biol.* **2013**, *9*, 297–299. [CrossRef]

40. Ogino, S.; Kubo, S.; Umemoto, R.; Huang, S.; Nishida, N.; Shimada, I. Observation of NMR signals from proteins introduced into living Mammalian cells by reversible membrane permeabilization using a pore-forming toxin, streptolysin o. *J. Am. Chem. Soc.* **2009**, *131*, 10834–10835. [CrossRef]

41. Sakakibara, D.; Sasaki, A.; Ikeya, T.; Hamatsu, J.; Hanashima, T.; Mishima, M.; Yoshimasu, M.; Hayashi, N.; Mikawa, T.; Waälchli, M.; et al. Protein structure determination in living cells by in-cell NMR spectroscopy. *Nature* **2009**, *458*, 102–105. [CrossRef]

42. Mobli, M.; Stern, A.S.; Hoch, J.C. Spectral reconstruction methods in fast NMR: Reduced dimensionality, random sampling and maximum entropy. *J. Magn. Reson.* **2006**, *182*, 96–105. [CrossRef]

43. Laue, E.D.; Mayger, M.R.; Skilling, J.; Staunton, J. Reconstruction of phase sensitive 2D NMR spectra by maximum entropy. *J. Magn. Reson.* **1986**, *68*, 14–29.

44. Hoch, J.C.; Maciejewski, M.W.; Mobli, M.; Schuyler, A.D.; Stern, A.S. Nonuniform Sampling and Maximum Entropy Reconstruction in Multidimensional NMR. *Acc. Chem. Res.* **2014**, *47*, 708–717. [CrossRef]

45. Ikeya, T.; Hanashima, T.; Hosoya, S.; Shimazaki, M.; Ikeda, S.; Mishima, M.; Güntert, P.; Ito, Y. Improved in-cell structure determination of proteins at near-physiological concentration. *Sci. Rep.* **2016**, *6*, 38312. [CrossRef] [PubMed]

46. Beck, B.D. Polymerization of the bacterial elongation factor for protein synthesis, EF-Tu. *Eur. J. Biochem.* **1979**, *97*, 495–502. [CrossRef] [PubMed]

47. Beck, M.; Schmidt, A.; Malmstroem, J.; Claassen, M.; Ori, A.; Szymborska, A.; Herzog, F.; Rinner, O.; Ellenberg, J.; Aebersold, R. The quantitative proteome of a human cell line. *Mol. Syst. Biol.* **2011**, *7*, 549. [CrossRef]

48. Boucher, W. *Azara, V2.0*; Department of Biochemistry, University of Cambridge: Cambridge, UK, 1996.

49. Schmidt, E.; Güntert, P. A new algorithm for reliable and general NMR resonance assignment. *J. Am. Chem. Soc.* **2012**, *134*, 12817–12829. [CrossRef] [PubMed]

50. Ikeya, T.; Jee, J.G.; Shigemitsu, Y.; Hamatsu, J.; Mishima, M.; Ito, Y.; Kainosho, M.; Güntert, P. Exclusively NOESY-based automated NMR assignment and structure determination of proteins. *J. Biomol. NMR* **2011**, *50*, 137–146. [CrossRef]

51. Schmidt, E.; Güntert, P. Reliability of exclusively NOESY-based automated resonance assignment and structure determination of proteins. *J. Biomol. NMR* **2013**, *57*, 193–204. [CrossRef]

52. Pritišanac, I.; Würz, J.M.; Alderson, T.R.; Güntert, P. Automatic structure-based NMR methyl resonance assignment in large proteins. *bioRxiv* **2019**, *1*, 538272.

53. Ikeya, T.; Ikeda, S.; Kigawa, T.; Ito, Y.; Güntert, P. Protein NMR Structure Refinement based on Bayesian Inference. *J. Phys. Conf. Ser.* **2016**, *699*, 012005. [CrossRef]

54. Tanaka, T.; Ikeya, T.; Kamoshida, H.; Suemoto, Y.; Mishima, M.; Shirakawa, M.; Güntert, P.; Ito, Y. High Resolution Protein 3D Structure Determination in Living Eukaryotic Cells. *Angew. Chem. Int. Ed. Engl.* **2019**, *58*. [CrossRef]

55. Sharaf, N.G.; Barnes, C.O.; Charlton, L.M.; Young, G.B.; Pielak, G.J. A bioreactor for in-cell protein NMR. *J. Magn. Reson.* **2010**, *202*, 140–146. [CrossRef] [PubMed]

56. Kubo, S.; Nishida, N.; Udagawa, Y.; Takarada, O.; Ogino, S.; Shimada, I. A gel-encapsulated bioreactor system for NMR studies of protein-protein interactions in living mammalian cells. *Angew. Chem. Int. Ed. Engl.* **2013**, *52*, 1208–1211. [CrossRef] [PubMed]

57. Inomata, K.; Kamoshida, H.; Ikari, M.; Ito, Y.; Kigawa, T. Impact of cellular health conditions on the protein folding state in mammalian cells. *Chem. Commun. (Camb.)* **2017**, *53*, 11245–11248. [CrossRef]

58. Hyberts, S.G.; Takeuchi, K.; Wagner, G. Poisson-gap sampling and forward maximum entropy reconstruction for enhancing the resolution and sensitivity of protein NMR data. *J. Am. Chem. Soc.* **2010**, *132*, 2145–2147. [CrossRef]

59. Harada, R.; Tochio, N.; Kigawa, T.; Sugita, Y.; Feig, M. Reduced native state stability in crowded cellular environment due to protein-protein interactions. *J. Am. Chem. Soc.* **2013**, *135*, 3696–3701. [CrossRef] [PubMed]

60. Hembram, D.S.; Haremaki, T.; Hamatsu, J.; Inoue, J.; Kamoshida, H.; Ikeya, T.; Mishima, M.; Mikawa, T.; Hayashi, N.; Shirakawa, M.; et al. An in-cell NMR study of monitoring stress-induced increase of cytosolic Ca^{2+} concentration in HeLa cells. *Biochem. Biophys. Res. Commun.* **2013**, *438*, 653–659. [CrossRef] [PubMed]

61. Kainosho, M.; Torizawa, T.; Iwashita, Y.; Terauchi, T.; Mei Ono, A.; Güntert, P. Optimal isotope labelling for NMR protein structure determinations. *Nature* **2006**, *440*, 52–57. [CrossRef]

62. Kainosho, M.; Miyanoiri, Y.; Terauchi, T.; Takeda, M. Perspective: Next generation isotope-aided methods for protein NMR spectroscopy. *J. Biomol. NMR* **2018**, *71*, 119–127. [CrossRef] [PubMed]

63. Liu, D.; Xu, R.; Cowburn, D. Chapter 8 Segmental Isotopic Labeling of Proteins for Nuclear Magnetic Resonance. *Methods Enzymol.* **2009**, *462*, 151–175.

64. Xue, J.; Burz, D.S.; Shekhtman, A. Segmental labeling to study multidomain proteins. *Adv. Exp. Med. Biol.* **2012**, *992*, 17–33.

65. Minato, Y.; Ueda, T.; Machiyama, A.; Shimada, I.; Iwai, H. Segmental isotopic labeling of a 140 kDa dimeric multi-domain protein CheA from Escherichia coli by expressed protein ligation and protein trans-splicing. *J. Biomol. NMR* **2012**, *53*, 191–207. [CrossRef]

66. Freiburger, L.; Sonntag, M.; Hennig, J.; Li, J.; Zou, P.; Sattler, M. Efficient segmental isotope labeling of multi-domain proteins using Sortase A. *J. Biomol. NMR* **2015**, *63*, 1–8. [CrossRef] [PubMed]

67. Mikula, K.M.; Tascon, I.; Tommila, J.J.; Iwai, H. Segmental isotopic labeling of a single-domain globular protein without any refolding step by an asparaginyl endopeptidase. *FEBS Lett.* **2017**, *591*, 1285–1294. [CrossRef]

68. Moult, J.; Fidelis, K.; Kryshtafovych, A.; Schwede, T.; Tramontano, A. Critical assessment of methods of protein structure prediction (CASP)-Round XII. *Proteins* **2018**, *86*, 7–15. [CrossRef] [PubMed]

69. Holland, D.J.; Bostock, M.J.; Gladden, L.F.; Nietlispach, D. Fast multidimensional NMR spectroscopy using compressed sensing. *Angew. Chem. Int. Ed. Engl.* **2011**, *50*, 6548–6551. [CrossRef] [PubMed]

70. Kazimierczuk, K.; Orekhov, V.Y. Accelerated NMR spectroscopy by using compressed sensing. *Angew. Chem. Int. Ed. Engl.* **2011**, *50*, 5556–5559. [CrossRef]

71. Hyberts, S.G.; Milbradt, A.G.; Wagner, A.B.; Arthanari, H.; Wagner, G. Application of iterative soft thresholding for fast reconstruction of NMR data non-uniformly sampled with multidimensional Poisson Gap scheduling. *J. Biomol. NMR* **2012**, *52*, 315–327. [CrossRef]

72. Ying, J.; Delaglio, F.; Torchia, D.A.; Bax, A. Sparse multidimensional iterative lineshape-enhanced (SMILE) reconstruction of both non-uniformly sampled and conventional NMR data. *J. Biomol. NMR* **2017**, *68*, 101–118. [CrossRef]

73. Kobayashi, N.; Hattori, Y.; Nagata, T.; Shinya, S.; Güntert, P.; Kojima, C.; Fujiwara, T. Noise peak filtering in multi-dimensional NMR spectra using convolutional neural networks. *Bioinformatics* **2018**, *34*, 4300–4301. [CrossRef]

74. Yu, I.; Mori, T.; Ando, T.; Harada, R.; Jung, J.; Sugita, Y.; Feig, M. Biomolecular interactions modulate macromolecular structure and dynamics in atomistic model of a bacterial cytoplasm. *Elife* **2016**, *5*, e19274. [CrossRef] [PubMed]

International Journal of
Molecular Sciences

Article

In-Cell NMR Study of Tau and MARK2 Phosphorylated Tau

Shengnan Zhang [1,*,†], Chuchu Wang [1,2,†], Jinxia Lu [3], Xiaojuan Ma [1,2], Zhenying Liu [1,2], Dan Li [3], Zhijun Liu [4] and Cong Liu [1,*]

[1]   Interdisciplinary Research Center on Biology and Chemistry, Shanghai Institute of Organic Chemistry, Chinese Academy of Sciences, 26 Qiuyue Road, Shanghai 201210, China; wangcc@sioc.ac.cn (C.W.); maxiaojuan@sioc.ac.cn (X.M.); liuzy@sioc.ac.cn (Z.L.)
[2]   Interdisciplinary Research Center on Biology and Chemistry, Shanghai Institute of Organic Chemistry, University of the Chinese Academy of Sciences, 19 A Yuquan Road, Shijingshan District, Beijing 100049, China
[3]   Key Laboratory for the Genetics of Developmental and Neuropsychiatric Disorders (Ministry of Education), Bio-X Institutes, Shanghai Jiao Tong University, Shanghai 200030, China; lujx@sjtu.edu.cn (J.L.); lidan2017@sjtu.edu.cn (D.L.)
[4]   National Facility for Protein Science in Shanghai, Zhangjiang Lab, Shanghai Advanced Research Institute, Chinese Academy of Sciences, Shanghai 201210, China; liuzhijun@sari.ac.cn
*   Correspondence: zhangshengnan@sioc.ac.cn (S.Z.); liulab@sioc.ac.cn (C.L.)
†   These authors contributed equally to this work.

Received: 5 December 2018; Accepted: 22 December 2018; Published: 26 December 2018

Abstract: The intrinsically disordered protein, Tau, is abundant in neurons and contributes to the regulation of the microtubule (MT) and actin network, while its intracellular abnormal aggregation is closely associated with Alzheimer's disease. Here, using in-cell Nuclear Magnetic Resonance (NMR) spectroscopy, we investigated the conformations of two different isoforms of Tau, Tau40 and k19, in mammalian cells. Combined with immunofluorescence imaging and western blot analyses, we found that the isotope-enriched Tau, which was delivered into the cultured mammalian cells by electroporation, is partially colocalized with MT and actin filaments (F-actin). We acquired the NMR spectrum of Tau in human embryonic kidney 293 (HEK-293T) cells, and compared it with the NMR spectra of Tau added with MT, F-actin, and a variety of crowding agents, respectively. We found that the NMR spectrum of Tau in complex with MT best recapitulates the in-cell NMR spectrum of Tau, suggesting that Tau predominantly binds to MT at its MT-binding repeats in HEK-293T cells. Moreover, we found that disease-associated phosphorylation of Tau was immediately eliminated once phosphorylated Tau was delivered into HEK-293T cells, implying a potential cellular protection mechanism under stressful conditions. Collectively, the results of our study reveal that Tau utilizes its MT-binding repeats to bind MT in mammalian cells and highlight the potential of using in-cell NMR to study protein structures at the residue level in mammalian cells.

Keywords: in-cell NMR; Tau; MARK2 phosphorylation; mammalian cells

---

1. Introduction

The intracellular environment is highly crowded, viscous, and packed with different proteins, nuclear acids, lipids, and other biomolecules. The structure and function of a certain protein is defined by the interplay between the protein and its neighboring biomolecules in the intracellular environment [1–4]. Therefore, it is of great importance to study the atomic structure and conformational dynamics of proteins within living cells, especially for intrinsically disordered proteins that lack well-folded 3D structures in the absence of binding partners [5–7]. However, conventional structural biology methods—including X-ray crystallography, solution NMR, and Cryogenic Electron Microscopy (cryo-EM)—deal with proteins purified in vitro, either in an aqueous solution or in a crystal structure, where the environment is highly simplified and fundamentally distinct from that inside the cell. Notably, recently developed in-cell NMR spectroscopy has emerged as a powerful technique to characterize the structures of proteins of interest within living cells at an atomic level [2,3,8–11]. To date, in-cell NMR has been successfully developed to characterize the protein structure [12], dynamics [13], and interactions [14,15] in bacteria. However, in contrast to eukaryotic cells, prokaryotic cells only reveal limited biological activities and lack significant biological processes such as protein maturation and post-translational modification, which can alter the structure and function of proteins. Thus, much effort has been devoted to extending in-cell NMR studies in eukaryotic cells, where the most technically challenging work is in introducing selective isotope-labeled proteins of interest in cells for NMR detection. Microinjection into *Xenopus laevis* oocytes [16–18], endocytotic transportation mediated by a cell-penetrating peptide [1,19], and diffusion through pore-forming toxins [20] have already been developed to successfully deliver isotopically labeled proteins purified in vitro to eukaryotic cells. Most recently, electroporation was shown to be as an effective and general approach to deliver isotope-labeled proteins into different types of mammalian cells [6,21]. Therefore, advances in the methodology of in-cell NMR pave the way toward investigating the structures and conformational dynamics of different proteins in the intracellular environment.

Tau is a typical intrinsically disordered protein that is highly abundant in the central nervous system [22,23]. It is capable of binding to a variety of proteins and other biomolecules including MT, heparin, and lipid molecules [24–28]. The physiological function of Tau is involved in the regulation and stabilization of the MT and actin network [29–31]. Tau contains multiple sites for post-translational modifications (e.g., phosphorylation, acetylation, methylation, and ubiquitination) under different cellular conditions for either the regulation of its normal function or in the pathogenesis of a disease [32]. For instance, hyperphosphorylation of Tau leads to the detachment of Tau from MT into the cytosol and the formation of abnormal filamentous amyloid aggregates [33–35]. These filamentous aggregates are the pathological hallmarks of a variety of neurodegenerative diseases including Alzheimer's disease (AD) [36], Pick's disease [37], and progressive supranuclear palsy [38]. Human tau in the brain has six isoforms that range from 352 to 441 amino acids in length [39]. The six isoforms differ in the number of MT-binding repeats (three or four) and insertions in the N-terminal projection domain (zero, one, or two). Cryo-EM studies have revealed that the MT-binding repeats are composed of an amyloid fibril core of filamentous Tau aggregates isolated from patient brains [36,37]. In contrast to the intensive investigation on the aggregated forms of Tau formed under pathogenic conditions, the structural studies on the soluble form of Tau—especially the conformation of Tau in the intracellular environment, and its relationship with its physiological function—are very limited.

In this study, we investigated the structures of two different isoforms of Tau, Tau40 and k19, in mammalian cells using in-cell NMR spectroscopy. The isotopically labeled Tau proteins were efficiently delivered into HEK-293T cells by electroporation. In combination with immunofluorescence imaging and in vitro NMR titration experiments, we confirmed that Tau/k19 can bind to both MT and F-actin in vitro, and they partially colocalize with MT and F-actin in the mammalian cells. The solution NMR spectrum of k19 in complex with MT best recapitulates the in-cell NMR spectrum of k19, suggesting that k19 predominantly binds to MT in the HEK-293T cells. Moreover, we found that microtubule affinity-regulating kinase 2 (MARK2) phosphorylated k19 was immediately dephosphorylated once being delivered into the HEK-293T cells. Our study reveals that Tau utilizes its MT-binding repeats to bind MT in mammalian cells, and highlights the potential of using in-cell NMR to study protein structure at the residue level in mammalian cells.

2. Results

2.1. In-Cell NMR Study of Tau k19

We first sought to investigate the structure of the three MT-binding repeats of Tau–k19 in mammalian cells using in-cell NMR, since k19 with 98 residues is much easier to study by NMR compared to Tau40 with 441 residues. Moreover, k19 contains the major AD related phosphorylation sites, and consists of the core sequence of filamentous Tau aggregates that is highly related to the pathology of Tau to AD. ^{15}N-labeled k19 was overexpressed and purified from *Escherichia coli (E. coli)*, and was then delivered into the HEK-293T cells by using a recently developed electroporation-based protocol (Figure 1a). To ensure that k19 is effectively delivered into the cells, we optimized different variables in the protocol. We optimized the concentration of k19 incubated with cells for electroporation ranging from 50–500 µM. We also optimized the electroporation parameters including pulse voltage (1200, 1300, or 1400 V), pulse width (10 or 20 milliseconds (ms)), and number of pulses (1 or 2). Finally, with a pulse program of 1400 V, 20 milliseconds (ms) and 1 pulse, ~10% of the k19 could be delivered into HEK-293T cells with an initial incubation concentration of 200 µM, as revealed by western blot analysis (Figure 1b). The concentration of ^{15}N-labeled k19 in cells was only slightly deceased seven hours after electroporation (Figure 1b), indicating a long half-time of the protein in HEK-293T cells. Immunofluorescence imaging experiments further confirmed that the ^{15}N-labeled k19 was successfully delivered into cells. Moreover, we found that k19 is mainly distributed in the cytoplasm, and it partially colocalizes with both MT and F-actin (Figure 1c). This result indicates that k19 may interact with different components in the cell, in contrast to another amyloid protein α-synuclein (α-syn), which has been previously studied by in-cell NMR and was found to be evenly distributed in the cytosol but did not exhibit specific interactions with other proteins [6].

To acquire a better quality and reproducible 2D NMR spectrum of the electroporated ^{15}N-labeled k19 in HEK-293T cells, we optimized the Bruker standard SOFAST-HMQC pulse [40,41] with the delay time (D1) set to 0.29 s and the optimized ^{1}H shape pulse. The 2D NMR spectrum was collected with 80 scans and 1024 × 128 complex points for ^{1}H (14 ppm) and ^{15}N (24 ppm), respectively, which resulted in a total of one hour for data collection. The 2D NMR spectra of k19 in HEK-293T cells with 3 and 7 h recovery were collected and displayed (Figure 1d). Western blot analysis confirmed that the collected NMR signals were mainly derived from the electroporated k19 in cells without significant leaking of k19 into the culture medium (Figure 1b). The NMR spectrum of k19 in cells shows similar patterns of crosspeaks to that of pure k19 in aqueous solution (Figure 1d). However, most resonances exhibited varied signal broadening and attenuation, especially the crosspeaks of residues 306VQIVYK311 (PHF6) that were previously identified to be crucial for fibrous Tau aggregation [42], exhibits significant broadening signals beyond detection (Figure 1e). This result suggests that k19 may interact with certain binding partners in cells, which is consistent with our imaging data that k19 colocalizes with MT and F-actin. Of note, two additional crosspeaks appeared in the in-cell NMR spectra; these two resonances exhibited enhanced signal intensities with the increasing recovery time without chemical shift perturbations, which may have been caused by a post-translational modification (e.g., phosphorylation, acetylation) on a particular residue of k19 in the HEK-293T cells.

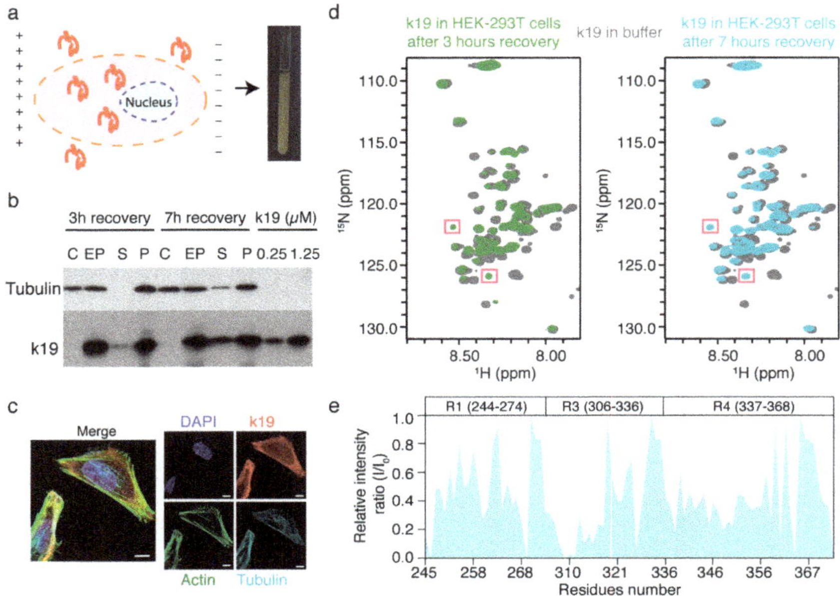

Figure 1. Characterization of the electroporated Tau k19 in mammalian cells. (**a**) Scheme of the electroporation procedure to deliver isotope-labeled k19 into mammalian cells and the resulting in-cell NMR sample. The orange and blue dotted circle stand for the cell and nucleus memberane of mammalian cells, respectively. + and − represent the electrical property of the electrode; (**b**) Western blot analysis of k19 electroporated into HEK-293T cells. Electroporated cells (EP) and untreated control cells (C) were diluted 10 times to be loaded with the same volume of purified k19 with indicated concentrations. S and P stand for the supernatant medium and cell pellet, respectively, of the in-cell NMR sample after NMR data collection; (**c**) Immunofluorescence detection of delivered k19 in SH-SY5Y cells after 7 h recovery. DAPI is used to stain nucleus. Scale bar, 7.5 μm; (**d**) Overlay of 2D in-cell NMR spectra of k19 in HEK-293T cells with 3 (green) and 7 (blue) hours recovery and in buffer (grey), respectively. The two additional crosspeaks in the in-cell NMR spectra are highlighted by the magenta boxes; (**e**) Residue-specific relative intensity ratio (I/I$_0$) of k19 in HEK-293T cells with 7 h recovery to that of k19 in buffer. Domain organization of k19 is illustrated on the top of the graph.

2.2. In Vitro NMR Characterization of Tau k19 with Crowding Agents

We next asked whether the crowding effect in the intracellular environment contributes to the specific pattern of signal broadening and attenuation of k19 observed in the in-cell NMR spectrum. To mimic the intracellular crowding environment, we selected four different types of crowding agents including bovine serum albumin (BSA, 200 g/L), Ficoll (200 g/L), lysozyme (10 g/L) and glycerol (20%: v/v) and measured the HSQC spectra of purified ^{15}N-labeled k19 in the presence of different agents, respectively. We found that addition of BSA and glycerol resulted in the global signal broadening of k19 (Figure 2a–c). On the contrary, addition with Ficoll and lysozyme caused significant enhancement of signal intensities for certain residues of k19, such as those around the sequence Lys-(Ile/Cys)-Gly-Ser (KXGS motif) of k19 (K257, S258, S262, T263, G323, S324, G326, and S352) and the subsequent sequence of Pro-Gly-Gly-Gly (PGGG motif, G271G272G273, G333G334G335, and G365G366G367N368), and moderate effects on the rest (Figure 2a,d,e). These results suggest that different crowding agents can influence the HSQC signal intensities of k19 in distinct ways. Whereas, none of these four crowding agents caused a similar intensity change profile as that of k19 in cells, indicating that the NMR spectrum of intracellular k19 may not be mainly due to crowding in the intracellular environment.

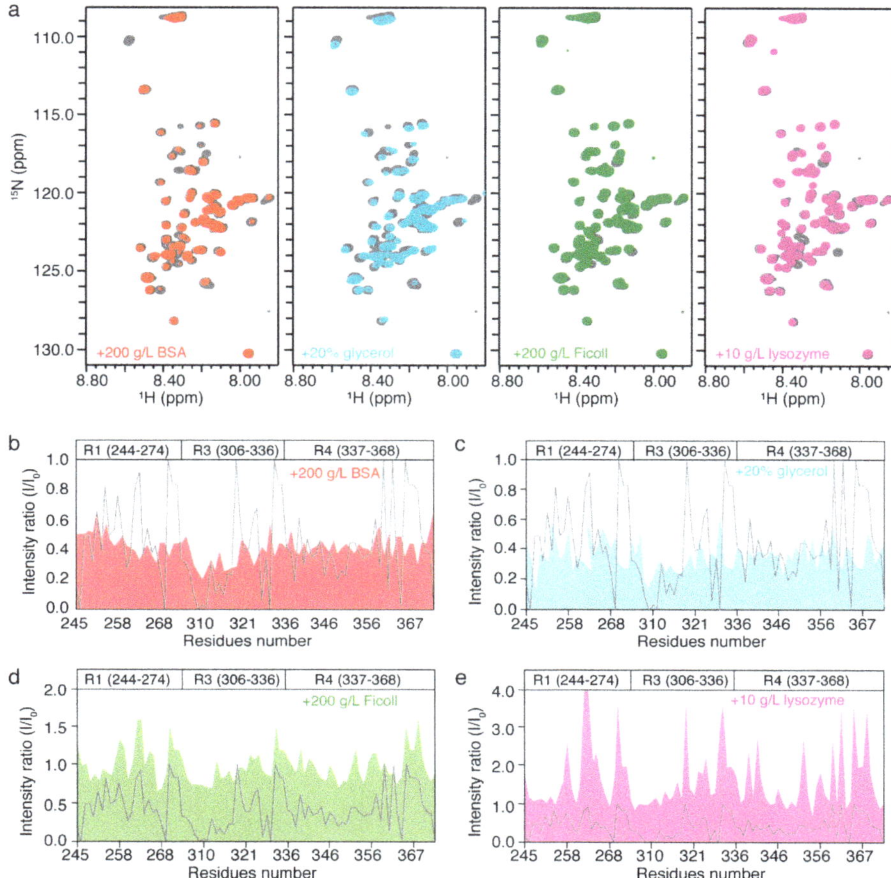

Figure 2. NMR characterization of k19 with crowding agents. (**a**) From left to right, overlay of the 2D ^{1}H-^{15}N HSQC spectra of 50 μM k19 in the absence (grey) and presence of 200 g/L BSA (red), 20% glycerol (blue), 200 g/L Ficoll (green), and 10 g/L lysozyme (magenta), respectively; (**b–e**) Residue-specific intensity ratio (I/I_0) of k19 in the presence of 200 g/L BSA (**b**), 20% glycerol (**c**), 200 g/L Ficoll (**d**), and 10 g/L lysozyme (**e**) to k19 in buffer. The relative intensity ratio (I/I_0) of k19 in HEK-293T cells to that of k19 in buffer is displayed as the grey line as a contrast.

2.3. In Vitro NMR Characterization of Tau k19 with MT and F-Actin

Tau protein was reported to bind and modulate the stability and dynamics of MT and F-actin, and can partially colocalize with both of them in cells based on our immunofluorescence imaging results. Thus, we asked how the binding of the MT and F-actin influence the characteristics of the k19 HSQC spectrum in vitro. We firstly prepared MT and F-actin from tubulin and actin, respectively, in vitro, and checked their formations by negative-staining transmission electron microscopy (TEM, Figure 3b,d). After centrifugation, the pellets of MT and F-actin were resuspended in NMR buffer and incubated with ^{15}N-labeled k19, respectively. As shown in Figure 3a, the addition of MT caused severe signal broadening of k19. Notably, the resulting spectrum resembled the major characteristics of that of k19 acquired in HEK-293T cells, especially the PHF6 region, which exhibited significant broadening signals beyond detection. However, addition of MT resulted in a more severe signal intensity reduction of the N-terminal region of R1, which is the most dramatic difference between the two spectra (Figure 3b). In contrast, addition of F-actin results in the global decrease of NMR signals all over the k19 sequence. While, the residues around the KXGS (K257, S258, S262, T263, G323, G326, and S356) and PGGG (G271-G273, G333-G335, and G365-T369) motifs of k19 exhibit enhancement of the HSQC signal intensities (Figure 3c,d). Together, our results demonstrate that k19 can interact with both MT and F-actin in vitro through distinct binding interfaces. Moreover, the spectrum of k19 in complex with MT fits well with the in-cell NMR spectrum of k19, indicating that electroporated ^{15}N-labeled k19 may predominantly bind to MT via its PHF6 region in HEK-293T cells.

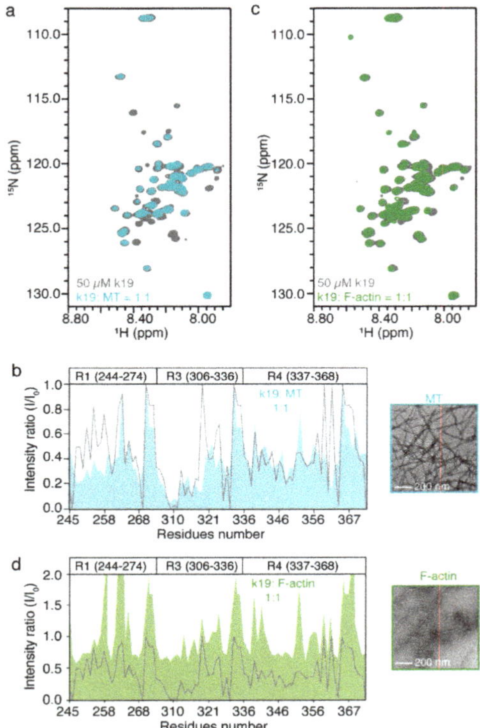

Figure 3. NMR characterization of k19 with microtubule (MT) and F-actin in vitro. (**a,c**) Overlay of the 2D ^{1}H-^{15}N HSQC spectra of 50 μM k19 in the absence (grey) or presence of MT (**a, blue**) and F-actin (**c, green**) at a molar ratio of 1:1, respectively; (**b,d**) Residue-specific intensity ratio (I/I_0) of k19 in the presence of MT (**b**) and F-actin (**d**) to that of k19 in buffer, respectively. The relative intensity ratio (I/I_0) of k19 in HEK-293T cells to that of k19 in buffer is displayed as a grey line as a contrast. TEM images of MT and F-actin used in the NMR sample are displayed on the right.

2.4. In-Cell NMR of MARK2 Phosphorylated k19

Hyperphosphorylation of Tau by microtubule affinity regulating kinase 2 (MARK2) was reported as a key event in disrupting the MT binding of Tau and promoting its abnormal aggregation, which is closely associated with Alzheimer's disease (AD) [35]. We next asked how the disease-associated MARK2-phosphorylated Tau (pk19) behaves in HEK-293T cells [33,35]. We first prepared the pk19 in vitro. Previous studies showed that MARK2 could phosphorylate the serine residues in the KXGS motifs of Tau's repeat domain [43]. Consistently, we found that MARK2 can phosphorylate k19 at S262, S324, S352, and S356, as shown in the HSQC spectrum with the crosspeaks of these four residues featuring a significant downfield shift caused by phosphorylation (Figure 4a). We then electroporated the ^{15}N-labeled pk19 into the HEK-293T cells and acquired the in-cell NMR spectrum. Intriguingly, the 2D NMR spectrum of pk19 in HEK-293T cells was the same as that of k19 in cells (Figure 4b). The phosphorylation of k19 on the four serine residues by MARK2 was eliminated once the pk19 protein was delivered into the cells. Western blot analysis further confirmed that only unphosphorylated k19 but not pk19 (anti-phosphorylated S356 Tau) could be detected in cells right after electroporation (Figure 4c). Moreover, after being dephosphorylated, delivered pk19 also distributed in the cytoplasm, and partially colocalized with both MT and F-actin (Figure 4d). Together, our data demonstrate that the disease-related pk19 was immediately dephosphorylated after being delivered into the HEK-293T cells, suggesting that the HEK-293T cells may utilize an unknown mechanism to protect Tau from being phosphorylated by MARK2 under normal conditions.

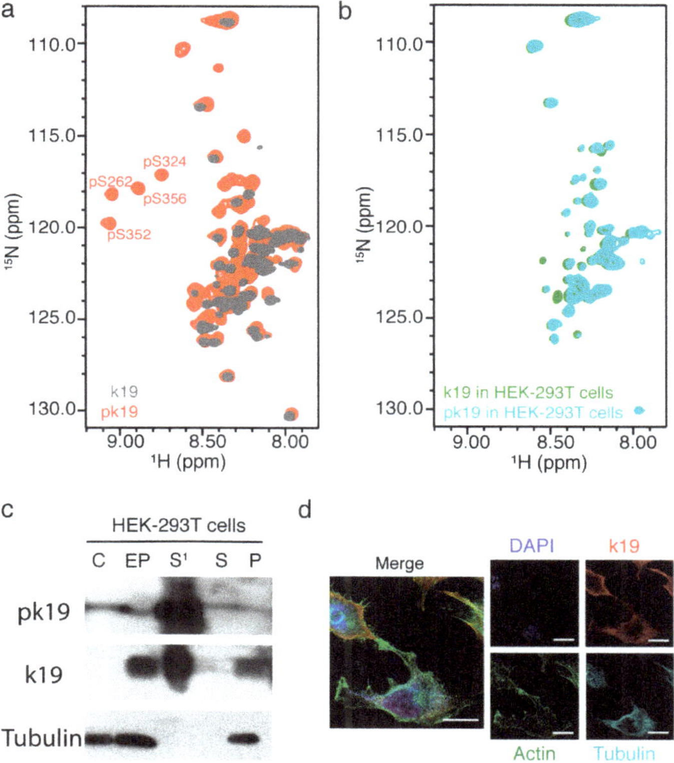

Figure 4. Characterization of the electroporated pk19 in mammalian cells. (**a**) Overlay of the 2D ^{1}H-^{15}N HSQC spectra of MARK2 phosphorylated k19 (pk19, red) and k19 (grey). The four phosphorylated serine residues including pS262, pS324, pS352, and pS356 are labeled. (**b**) Overlay of 2D in-cell NMR spectra of pk19 (blue) and k19 (green) in HEK-293T cells; (**c**) Western blot analysis of pk19 electroporated into HEK-293T cells. C, EP, S^{1}, S, and P stand for untreated control cells, pk19 electroporated cells, supernatant medium of the cell–protein mixture after electroporation, supernatant medium and cell pellet of the in-cell NMR sample after NMR experiment collection, respectively. The antibody used for pk19 can only detect Tau when phosphorylated at S356, while the antibody for k19 can recognize both non-phosphorylated and phosphorylated S262 of Tau; (**d**) Immunofluorescence detection of delivered pk19 in SH-SY5Y cells after seven hours' recovery. Scale bar, 7.5 μm. The antibody used here can recognize both non-phosphorylated and phosphorylated S262 of Tau.

2.5. In-Cell NMR of Full-Length Tau40

Finally, we sought to acquire the in-cell NMR spectrum of the largest isoform of Tau—Tau40 which contains 441 amino acids (Figure 5a). Firstly, we purified the ^{15}N-labeled Tau40 in vitro and collected its 2D HSQC spectrum in aqueous solution (Figure 5b). The spectrum revealed a narrow and highly congested cluster of the amide proton signals, indicating that Tau40 adopts an intrinsically disordered conformation in an aqueous solution, which is consistent with previous studies [44]. Next, we electroporated the ^{15}N-labeled Tau40 into the HEK-293T cells. The 2D in-cell NMR spectrum showed that a lot of resonances suffered severe signal broadening (Figure 5a), indicating that Tau40 binds to potential partners in cells. Moreover, similar to the in-cell NMR spectrum of k19, we found that most residues in the MT-binding repeats of Tau40 revealed significant signal intensity reduction, especially the residues around the PHF6 region. The signals of residues V309, Y310, and K311 in the in-cell NMR spectrum of Tau40 was completely unobservable, suggesting that these residues might bind to MT which is consistent with that of k19. Immunofluorescence imaging further showed that the delivered Tau40 was predominantly distributed in the cytoplasm and partially colocalized with MT and F-actin which is similar to electroporated k19 (Figure 5c). Taken together, our results reveal that Tau40 utilizes its MT-binding repeats to bind MT in mammalian cells.

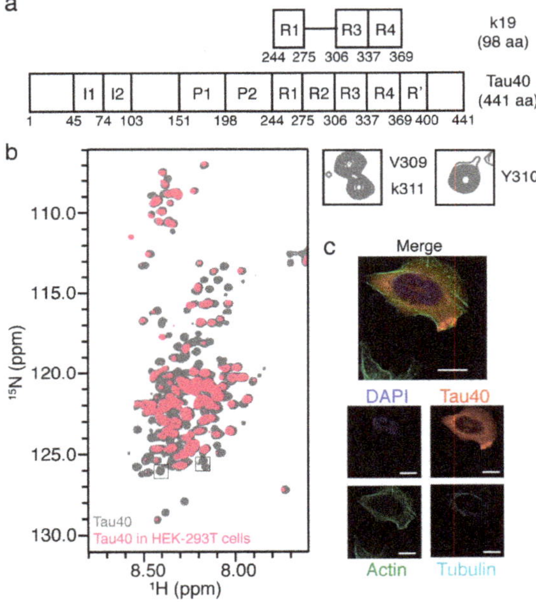

Figure 5. Characterization of the electroporated Tau40 in mammalian cells. (**a**) Bar diagram of full-length Tau40 and the fragment of k19. I1 and I2 stand for the two insertions in the N-terminal projection domain. The four MT-binding repeats are donated as R1-R4, while P1 and P2 represent the basic and proline-rich region preceding the repeats. R' is the C-terminal flanking region; (**b**) Overlay of 2D in-cell NMR spectra of Tau40 in HEK-293T cells and in buffer (grey), the three crosspeaks of residues in the PHF6 region including V309, Y310, and Y311 are zoomed in on the right; (**c**) Immunofluorescence detection of delivered Tau40 in SH-SY5Y cells after seven hours' recovery. Scale bar, 7.5 μm.

3. Discussion

As introduced in 1975 [45], the in-cell NMR method has greatly expanded the toolbox of structural biology to explore the structure and dynamics of proteins in cells. Particularly, recent in-cell NMR studies on α-syn elegantly demonstrated that α-syn adopts as a monomer with a compacted disordered conformation in the cellular environment [6], and methionine-oxidized α-syn at the N-terminal, but not at the C-terminal, can be repaired by an intracellular reductase, revealing the general mechanism of site-selective α-syn repair in cells [21]. Herein, we utilized the in-cell NMR method to investigate the structure of the two isoforms of Tau, k19 and Tau40, in HEK-293T cells. We found that exogenously delivered k19 and Tau40 remain largely disordered in cells. Moreover, the MT-binding repeats of both two isoforms can bind to MT and F-actin in cells. Intriguingly, we found two additional unidentified crosspeaks for the delivered k19 in cells implying that post-translational modification may have occurred once k19 was delivered into the HEK-293T cells. However, the positions of these two crosspeaks in the NMR spectrum did not match with the phosphorylated sites studied previously by NMR in vitro [46–49]. A previous in-cell NMR study revealed that α-syn is acetylated at the N-terminal once electroporated into mammalian cells. Therefore, Tau might be acetylated or modified by other post-translational modifications such as methylation or phosphorylation in mammalian cells; this will be studied through combining protein mass spectrometry in the future. Notably, we observed that the exogenously delivered pk19, which is closely associated with disease, was immediately dephosphorylated in HEK-293T cells. This result implies that, in a normal cellular environment, the cell is engaged with a sophisticated post-translational modification system that safeguards the proper post-translational modification on different proteins and corrects abnormal modifications—in this case, MARK2-mediated phosphorylation on k19. Dysregulation of this system upon aging or under disease conditions may lead to hyperphosphorylation of Tau and consequently to abnormal aggregation.

We noticed that the overall 2D HSQC spectrum of electroporated Tau40 in the HEK-293T cells was similar to the previously acquired spectrum of Tau40 in the *Xenopus laevis* oocytes, delivered using microinjection [16]. However, we did not observe the additional resonances for Tau40 in HEK-293T cells which was previously identified as a possible phosphorylation resonance of Tau40 modified in *Xenopus laevis* oocytes. A recent in-cell NMR study revealed that cell type specifically contributes to the biological and pathological behaviors of α-syn in different intracellular environments [6,50]. Therefore, different cell types may feature distinct intracellular environment as well as distinct post-translational modification systems. Further in-cell NMR studies of Tau in neuron or neuronal-like cells may prove to be of great importance in revealing the physiological in-cell structure and dynamics of Tau.

4. Materials and Methods

4.1. Protein Overexpression and Purification

Human Tau40/k19 was expressed and purified on the basis of a previously described method [51]. Briefly, expression of Tau40/k19 was induced in *E. coli* BL21 (DE3) by addition of 0.5 mM isopropyl β-D-1-thiogalactopyranoside (IPTG) with the OD_{600} of 0.4–0.6 and grew overnight at 20 °C. After harvesting the cells, Tau40/k19 was purified by a HighTrap HP SP (5 mL) column (GE Healthcare, *Chicago*, IL, USA), followed by a Superdex 75 gel filtration column (GE Healthcare, *Chicago*, IL, USA). The final purified protein was buffer exchanged to NMR buffer (pH 6.5, 50 mM sodium phosphate buffer (PBS) and 50 mM NaCl), concentrated, and stored at −80 °C. Protein concentration was determined bybicinchoninic acid(BCA) assay (Thermo Fisher Scientific, Waltham, MA, USA).

For ^{15}N- or $^{15}N/^{13}C$-labeled proteins, protein expression and purification were the same as that for unlabeled proteins except that the cells were grown in M9 minimal medium with $^{15}NH_4Cl$ (1 g/L) in the absence or presence of $^{13}C_6$-glucose (2 g/L).

4.2. In Vitro k19 Phosphorylation

Phosphorylation of k19 by MARK2 kinase was carried out following a method described previously [47]. Briefly, k19 was incubated with a hyperactive variant (T208E) of cat MARK2 [52] at a molar ratio of 10:1 in a buffer of 50 mM Hepes, pH 8.0, 150 mM KCl, 10 mM $MgCl_2$, 5 mM ethylene glycol tetraacetic acid (EGTA), 1 mM phenylmethylsulfonyl fluoride (PMSF), 1 mM dithiothreitol

(DTT), 2 mM ATP (Sigma, Saint Louis, MO, USA), and protease inhibitor cocktail (Roche, Basel, Swizerland) at 30 °C overnight. Phosphorylated protein was further purified by high-performance liquid chromatography (HPLC) (Agilent, Santa Clara, CA, USA) to remove kinase, and lyophilized. The sites and degrees of phosphorylation were quantified using 2D ^{1}H-^{15}N HSQC spectrum according to the previously publications [47,53].

4.3. Electroporation of Purified Proteins into Mammalian Cells

Human HEK-293T (ATCC, CRL-3216) and SH-SY5Y (ATCC, CRL-2266) cells were cultured following the methods provided by ATCC (Manassas, VA, USA). Both cell lines are routinely tested for mycoplasma contaminations and are mycoplasma free.

In-cell NMR samples were prepared using the modified protocol according to the previous publication [6]. The cells were passaged about 4–6 times prior to NMR experiments. The cells were collected by trypsinization and washed with PBS three times. Purified Tau/k19/pk19 (~1 mM) was diluted to a final concentration of 200 μM with Buffer R supplied in the Neon transfection system kit (Invitrogen, MPK10025, Carlsbad, CA, USA). Then cells in PBS were pellet again and then resuspended with the Tau/k19/pk19 solution to a cell density of ~8 × 10^7 per mL. Electroporation was conducted using 100 μL of the cell-protein mixture with a pulse program of 1400 V (pulse voltage), 20 milliseconds (ms) (pulse width), and 1 pulse by the Neon transfection system (Invitrogen, MPK5000, Carlsbad, CA, USA).

For immunofluorescence and western blotting experiments, aliquots of 0.5 × 10^6 cells were added to separate wells of a 24-well plate, filled with 0.5 mL media. For in-cell NMR samples of k19, aliquots of 4~8 × 10^6 cells were added to eight 10-cm dishes with 10 mL media and cultured after 3~7 h for cell recovery. For pk19 and Tau40 samples, no recovery step was used. Then the cells were harvested and washed with PBS for four times. The in-cell NMR samples were taken up in pH-stable L-15 medium (Gibco, 11415064, Waltham, MA, USA) to a total volume of 500 μL with 10% D$_2$O and settled into the NMR tube.

4.4. Western Blot

To determine the delivered protein in cells and check the leakage of the in-cell NMR samples, each cell sample was centrifuged at 300 *g* for 3 min immediately after the experiment. Then the supernatants and pellets were resuspended in Laemmli buffer to a final volume of 500 μL, and boiled for 10 min. These samples were diluted by 10 times, and loaded to a 15% SDS-PAGE gel. The primary antibodies used were listed as follows: Tau/k19 (abcam, ab64193, 1:1000), pk19 (abcam, ab75603, 1:1000) and tubulin (abcam, ab7291, 1:1000). Delivered k19 in HEK-293T cells was ~25 μM measured by ImageJ (Available online: https://imagej.net/Welcome) [54].

4.5. Immunostaining of Cultured Cells

For immunofluorescence imaging of control and electroporated cells as fixed specimens, cells were recovered at 37 °C for 7–8 h on poly-D-lysine-coated coverslips in 24-well plates. Cells were washed by pre-warmed PBS three times to remove extracellular proteins, then fixed in 4% (*w/v*) paraformaldehyde in PBS for 30 min and permeabilized with 0.1% (*v/v*) Triton X-100 in PBS for 20 min. After washing with PBS for three times, cells were blocked with 10% BSA in PBS for 1 h followed by incubation of antibodies at 4 °C overnight, including anti-Tau (abcam, ab64193, 1:1000), anti-alpha tubulin (abcam, ab7291, 1:1000), and fluorescein isothiocyanate-labeled phalloidin (Yeasen, 40736ES75, 1:200) for F-actin along with cell membrane skeleton. Slides were washed 3 × 10 min with PBS and nuclei stained by antifade mountant coupled DAPI (Invitrogen, P36935, Carlsbad, CA, USA). The samples were observed by a confocal microscope (Lecia, SP8, Wetzlar, Germany).

4.6. MT Preparation

Tubulin (catalog no. T240, Cytoskeleton) was incubated at concentrations higher than 200 μM in microtubule assembly buffer (100 mM Pipes, pH 6.9, 1 mM EDTA, 1 mM $MgSO_4$, 1 mM DTT) in the presence of 1 mM GTP at 37 °C for 5 min. After addition of 100 μM paclitaxel (Sigma, Saint Louis, MO, USA), the polymerization was performed for 60 min at 37 °C. The integrity of MTs was checked by negative- staining TEM. Just before NMR experiments, samples were centrifuged at 14,000 rpm for 15 min, then the pellets were resuspended in NMR buffer.

4.7. F-Actin Preparation

Non-muscle human actin (catalog no. APHL99, Cytoskeleton) was first dissolved by 100 μL water to reach a concentration of 10 mg/mL and then diluted to 1 mg/mL with the actin buffer (5 mM Tris-HCl, pH 8.0, 0.2 mM $CaCl_2$, 0.2 mM ATP, 0.5 mM DTT, 5% (w/v) sucrose and 1% (w/v) dextran). After centrifugation at 14,000 rpm for 15 min, the supernatant was kept. Then actin was polymerized by addition of polymerization buffer (1/10th the volume; 100 mM Tris-HCl, pH 7.5, 20 mM $MgCl_2$, 500 mM KCl and 10 mM ATP) at room temperature for 1 h. The integrity of F-actins was checked by negative-staining TEM. Just before the NMR experiments, samples were spun down at 14,000 rpm for 15 min, then the pellets were resuspended in NMR buffer.

4.8. Transmission Electron Microscopy (TEM)

TEM images were collected on a Tecnai G2 Spirit TEM operated at an accelerating voltage of 120 kV. Samples (8 μL) were deposited on carbon-coated grids for 45 s. The grids were then washed twice with ddH_2O (8 μL) and incubated with 8 μL uranyl acetate (2%, v/v) for staining. Images were recorded using a 4K × 4K charge-coupled device camera (BM-Eagle, FEI Tecnai, Waltham, MA, USA).

4.9. In-Cell and In Vitro NMR Spectroscopy

All the NMR experiments were carried out at 25 °C on a Bruker 900 or 600 MHz spectrometer equipped with a cryogenic probe. The buffer used in all in vitro NMR assays was 50 mM sodium phosphate buffer (pH 6.5) containing 50 mM NaCl and 10% D_2O (v/v). Backbone resonance assignment of k19 and pk19 was accomplished based on the collected 3D HNCACB and CBCACONH spectra and assignments from previous studies [47,53].

For the in-cell NMR samples, Bruker standard SOFAST-HMQC pulse sequence [40,41] was used with the delay time (D1) of 0.29 s and the ^{1}H shape pulse efficient was optimized for collecting the 2D NMR spectrum. 1024 × 128 complex points were used for ^{1}H (14 ppm) and ^{15}N (24 ppm), respectively. The 2D SOFAST-HMQC spectrum was collected with 80 scans resulting in totally one hour experiment time. To calculate the residue-resolved relative intensity ratio (Y) of k19 in cells (I) versus in buffer (I_0) in Figure 1e, the equation $Y = [I(X)/I_0(X)]/[I(333)/I_0(333)]$ was used for each residue X, where the intensity ratio of residue X was normalized by that of the highly flexible residue G333.

For in vitro titration assays, each NMR sample was freshly prepared from high concentration protein stocks with a total volume of 500 μL (10% D_2O). Each 2D ^{1}H-^{15}N HSQC spectrum was collected with 16 scans per transient and complex points of 2048 × 160 for ^{1}H (14 ppm) and ^{15}N (24 ppm), respectively. All NMR data were processed by NMRpipe [55] and analyzed by NMRViewJ [56].

Author Contributions: Conceptualization, S.Z., C.W. and C.L.; Methodology, S.Z., C.W. and Z.L. (Zhijun Liu); Software, S.Z. and C.W.; Validation, S.Z. and C.W.; Formal Analysis, S.Z. and C.W.; Investigation, S.Z. and C.L.; Resources, J.L., X.M. and Z.L. (Zhenying Liu); Writing-Original Draft Preparation, S.Z.; Writing-Review & Editing, D.L. and C.L.; Visualization, S.Z. and C.W.; Supervision, S.Z. and C.L.; Project Administration, S.Z. and C.L.; Funding Acquisition, C.L.

Funding: This work was supported by the Major State Basic Research Development Program (grant no. 2016YFA0501902), the National Natural Science Foundation (NSF) of China (grant no. 31470748), the "1000 Talents Plan" of China.

Acknowledgments: We thank Songzi Jiang and other staff members of National Facility for Protein Science in Shanghai for assistance in NMR data collection.

Conflicts of Interest: The authors declare no conflict of interest.

References

1. Inomata, K.; Ohno, A.; Tochio, H.; Isogai, S.; Tenno, T.; Nakase, I.; Takeuchi, T.; Futaki, S.; Ito, Y.; Hiroaki, H.; et al. High-resolution multi-dimensional NMR spectroscopy of proteins in human cells. *Nature* **2009**, *458*, 106–109. [CrossRef] [PubMed]

2. Plitzko, J.M.; Schuler, B.; Selenko, P. Structural biology outside the box-inside the cell. *Curr. Opin. Struct. Biol.* **2017**, *46*, 110–121. [CrossRef] [PubMed]

3. Luchinat, E.; Banci, L. A unique tool for cellular structural biology: In-cell NMR. *J. Biol. Chem.* **2016**, *291*, 3776–3784. [CrossRef] [PubMed]

4. Bertrand, K.; Reverdatto, S.; Burz, D.S.; Zitomer, R.; Shekhtman, A. Structure of proteins in eukaryotic compartments. *J. Am. Chem. Soc.* **2012**, *134*, 12798–12806. [CrossRef] [PubMed]

5. Theillet, F.X.; Binolfi, A.; Frembgen-Kesner, T.; Hingorani, K.; Sarkar, M.; Kyne, C.; Li, C.; Crowley, P.B.; Gierasch, L.; Pielak, G.J.; et al. Physicochemical properties of cells and their effects on intrinsically disordered proteins (IDPs). *Chem. Rev.* **2014**, *114*, 6661–6714. [CrossRef] [PubMed]

6. Theillet, F.X.; Binolfi, A.; Bekei, B.; Martorana, A.; Rose, H.M.; Stuiver, M.; Verzini, S.; Lorenz, D.; van Rossum, M.; Goldfarb, D.; et al. Structural disorder of monomeric alpha-synuclein persists in mammalian cells. *Nature* **2016**, *530*, 45–50. [CrossRef]

7. Sciolino, N.; Burz, D.S.; Shekhtman, A. In-Cell NMR spectroscopy of intrinsically disordered proteins. *Proteomics* **2018**. [CrossRef]

8. Luchinat, E.; Banci, L. In-cell NMR: A topical review. *IUCrJ* **2017**, *4*, 108–118. [CrossRef]

9. Rahman, S.; Byun, Y.; Hassan, M.I.; Kim, J.; Kumar, V. Towards understanding cellular structure biology: In-cell NMR. *Biochim. Biophys. Acta Proteins Proteom.* **2017**, *1865*, 547–557. [CrossRef]

10. Burz, D.S.; Shekhtman, A. Structural biology: Inside the living cell. *Nature* **2009**, *458*, 37–38. [CrossRef]

11. Maldonado, A.Y.; Burz, D.S.; Shekhtman, A. In-cell NMR spectroscopy. *Prog. Nucl. Magn. Reson. Spectrosc.* **2011**, *59*, 197–212. [CrossRef] [PubMed]

12. Sakakibara, D.; Sasaki, A.; Ikeya, T.; Hamatsu, J.; Hanashima, T.; Mishima, M.; Yoshimasu, M.; Hayashi, N.; Mikawa, T.; Walchli, M.; et al. Protein structure determination in living cells by in-cell NMR spectroscopy. *Nature* **2009**, *458*, 102–105. [CrossRef] [PubMed]

13. Monteith, W.B.; Pielak, G.J. Residue level quantification of protein stability in living cells. *Proc. Natl. Acad. Sci. USA* **2014**, *111*, 11335–11340. [CrossRef] [PubMed]

14. Burz, D.S.; Dutta, K.; Cowburn, D.; Shekhtman, A. In-cell NMR for protein-protein interactions (STINT-NMR). *Nat. Protoc.* **2006**, *1*, 146–152. [CrossRef] [PubMed]

15. Burz, D.S.; Dutta, K.; Cowburn, D.; Shekhtman, A. Mapping structural interactions using in-cell NMR spectroscopy (STINT-NMR). *Nat. Methods* **2006**, *3*, 91–93. [CrossRef] [PubMed]

16. Selenko, P.; Frueh, D.P.; Elsaesser, S.J.; Haas, W.; Gygi, S.P.; Wagner, G. In situ observation of protein phosphorylation by high-resolution NMR spectroscopy. *Nat. Struct. Mol. Biol.* **2008**, *15*, 321–329. [CrossRef] [PubMed]

17. Selenko, P.; Serber, Z.; Gadea, B.; Ruderman, J.; Wagner, G. Quantitative NMR analysis of the protein G B1 domain in Xenopus laevis egg extracts and intact oocytes. *Proc. Natl. Acad. Sci. USA* **2006**, *103*, 11904–11909. [CrossRef] [PubMed]

18. Amata, I.; Maffei, M.; Igea, A.; Gay, M.; Vilaseca, M.; Nebreda, A.R.; Pons, M. Multi-phosphorylation of the intrinsically disordered unique domain of c-Src studied by in-cell and real-time NMR spectroscopy. *ChemBioChem* **2013**, *14*, 1820–1827. [CrossRef] [PubMed]

19. Danielsson, J.; Inomata, K.; Murayama, S.; Tochio, H.; Lang, L.; Shirakawa, M.; Oliveberg, M. Pruning the ALS-associated protein SOD1 for in-cell NMR. *J. Am. Chem. Soc.* **2013**, *135*, 10266–10269. [CrossRef] [PubMed]

20. Ogino, S.; Kubo, S.; Umemoto, R.; Huang, S.; Nishida, N.; Shimada, I. Observation of NMR signals from proteins introduced into living mammalian cells by reversible membrane permeabilization using a pore-forming toxin, streptolysin O. *J. Am. Chem. Soc.* **2009**, *131*, 10834–10835. [CrossRef]

21. Binolfi, A.; Limatola, A.; Verzini, S.; Kosten, J.; Theillet, F.X.; Rose, H.M.; Bekei, B.; Stuiver, M.; van Rossum, M.; Selenko, P. Intracellular repair of oxidation-damaged alpha-synuclein fails to target C-terminal modification sites. *Nat. Commun.* **2016**, *7*, 10251. [CrossRef] [PubMed]

22. Hirokawa, N.; Funakoshi, T.; Sato-Harada, R.; Kanai, Y. Selective stabilization of tau in axons and microtubule-associated protein 2C in cell bodies and dendrites contributes to polarized localization of cytoskeletal proteins in mature neurons. *J. Cell Biol.* **1996**, *132*, 667–679. [CrossRef] [PubMed]

23. Konzack, S.; Thies, E.; Marx, A.; Mandelkow, E.M.; Mandelkow, E. Swimming against the tide: Mobility of the microtubule-associated protein tau in neurons. *J. Neurosci.* **2007**, *27*, 9916–9927. [CrossRef] [PubMed]

24. Cabrales Fontela, Y.; Kadavath, H.; Biernat, J.; Riedel, D.; Mandelkow, E.; Zweckstetter, M. Multivalent cross-linking of actin filaments and microtubules through the microtubule-associated protein Tau. *Nat. Commun.* **2017**, *8*, 1981. [CrossRef] [PubMed]

25. Jones, E.M.; Dubey, M.; Camp, P.J.; Vernon, B.C.; Biernat, J.; Mandelkow, E.; Majewski, J.; Chi, E.Y. Interaction of tau protein with model lipid membranes induces tau structural compaction and membrane disruption. *Biochemistry* **2012**, *51*, 2539–2550. [CrossRef] [PubMed]

26. Mukrasch, M.D.; von Bergen, M.; Biernat, J.; Fischer, D.; Griesinger, C.; Mandelkow, E.; Zweckstetter, M. The "jaws" of the tau-microtubule interaction. *J. Biol. Chem.* **2007**, *282*, 12230–12239. [CrossRef] [PubMed]

27. Sibille, N.; Sillen, A.; Leroy, A.; Wieruszeski, J.M.; Mulloy, B.; Landrieu, I.; Lippens, G. Structural impact of heparin binding to full-length Tau as studied by NMR spectroscopy. *Biochemistry* **2006**, *45*, 12560–12572. [CrossRef]

28. Drechsel, D.N.; Hyman, A.A.; Cobb, M.H.; Kirschner, M.W. Modulation of the dynamic instability of tubulin assembly by the microtubule-associated protein tau. *Mol. Biol. Cell* **1992**, *3*, 1141–1154. [CrossRef]

29. Mandelkow, E.; Mandelkow, E.M. Microtubules and microtubule-associated proteins. *Curr. Opin. Cell Biol.* **1995**, *7*, 72–81. [CrossRef]

30. Elie, A.; Prezel, E.; Guerin, C.; Denarier, E.; Ramirez-Rios, S.; Serre, L.; Andrieux, A.; Fourest-Lieuvin, A.; Blanchoin, L.; Arnal, I. Tau co-organizes dynamic microtubule and actin networks. *Sci. Rep.* **2015**, *5*, 9964. [CrossRef]

31. Cunningham, C.C.; Leclerc, N.; Flanagan, L.A.; Lu, M.; Janmey, P.A.; Kosik, K.S. Microtubule-associated protein 2c reorganizes both microtubules and microfilaments into distinct cytological structures in an actin-binding protein-280-deficient melanoma cell line. *J. Cell Biol.* **1997**, *136*, 845–857. [CrossRef] [PubMed]

32. Martin, L.; Latypova, X.; Terro, F. Post-translational modifications of tau protein: Implications for Alzheimer's disease. *Neurochem. Int.* **2011**, *58*, 458–471. [CrossRef] [PubMed]

33. Ando, K.; Oka, M.; Ohtake, Y.; Hayashishita, M.; Shimizu, S.; Hisanaga, S.; Iijima, K.M. Tau phosphorylation at Alzheimer's disease-related Ser356 contributes to tau stabilization when PAR-1/MARK activity is elevated. *Biochem. Biophys. Res. Commun.* **2016**, *478*, 929–934. [CrossRef] [PubMed]

34. Biernat, J.; Gustke, N.; Drewes, G.; Mandelkow, E.M.; Mandelkow, E. Phosphorylation of Ser262 strongly reduces binding of tau to microtubules: Distinction between PHF-like immunoreactivity and microtubule binding. *Neuron* **1993**, *11*, 153–163. [CrossRef]

35. Gu, G.J.; Wu, D.; Lund, H.; Sunnemark, D.; Kvist, A.J.; Milner, R.; Eckersley, S.; Nilsson, L.N.; Agerman, K.; Landegren, U.; et al. Elevated MARK2-dependent phosphorylation of Tau in Alzheimer's disease. *J. Alzheimers Dis.* **2013**, *33*, 699–713. [CrossRef] [PubMed]

36. Fitzpatrick, A.W.P.; Falcon, B.; He, S.; Murzin, A.G.; Murshudov, G.; Garringer, H.J.; Crowther, R.A.; Ghetti, B.; Goedert, M.; Scheres, S.H.W. Cryo-EM structures of tau filaments from Alzheimer's disease. *Nature* **2017**, *547*, 185–190. [CrossRef] [PubMed]

37. Falcon, B.; Zhang, W.; Murzin, A.G.; Murshudov, G.; Garringer, H.J.; Vidal, R.; Crowther, R.A.; Ghetti, B.; Scheres, S.H.W.; Goedert, M. Structures of filaments from Pick's disease reveal a novel tau protein fold. *Nature* **2018**, *561*, 137–140. [CrossRef]

38. Flament, S.; Delacourte, A.; Verny, M.; Hauw, J.J.; Javoy-Agid, F. Abnormal Tau proteins in progressive supranuclear palsy. Similarities and differences with the neurofibrillary degeneration of the Alzheimer type. *Acta Neuropathol.* **1991**, *81*, 591–596. [CrossRef]

39. Goedert, M.; Spillantini, M.G.; Jakes, R.; Rutherford, D.; Crowther, R.A. Multiple isoforms of human microtubule-associated protein tau: Sequences and localization in neurofibrillary tangles of Alzheimer's disease. *Neuron* **1989**, *3*, 519–526. [CrossRef]

40. Schanda, P.; Brutscher, B. Very fast two-dimensional NMR spectroscopy for real-time investigation of dynamic events in proteins on the time scale of seconds. *J. Am. Chem. Soc.* **2005**, *127*, 8014–8015. [CrossRef]

41. Schanda, P.; Kupce, E.; Brutscher, B. SOFAST-HMQC experiments for recording two-dimensional heteronuclear correlation spectra of proteins within a few seconds. *J. Biomol. NMR* **2005**, *33*, 199–211. [CrossRef] [PubMed]

42. von Bergen, M.; Friedhoff, P.; Biernat, J.; Heberle, J.; Mandelkow, E.M.; Mandelkow, E. Assembly of tau protein into Alzheimer paired helical filaments depends on a local sequence motif ((306)VQIVYK(311)) forming beta structure. *Proc. Natl. Acad. Sci. USA* **2000**, *97*, 5129–5134. [CrossRef] [PubMed]

43. Yoshida, H.; Goedert, M. Phosphorylation of microtubule-associated protein tau by AMPK-related kinases. *J. Neurochem.* **2012**, *120*, 165–176. [CrossRef] [PubMed]

44. Narayanan, R.L.; Durr, U.H.; Bibow, S.; Biernat, J.; Mandelkow, E.; Zweckstetter, M. Automatic assignment of the intrinsically disordered protein Tau with 441-residues. *J. Am. Chem. Soc.* **2010**, *132*, 11906–11907. [CrossRef] [PubMed]

45. London, R.E.; Gregg, C.T.; Matwiyoff, N.A. Nuclear magnetic resonance of rotational mobility of mouse hemoglobin labeled with (2-13C)histidine. *Science* **1975**, *188*, 266–268. [CrossRef] [PubMed]

46. Leroy, A.; Landrieu, I.; Huvent, I.; Legrand, D.; Codeville, B.; Wieruszeski, J.M.; Lippens, G. Spectroscopic studies of GSK3{beta} phosphorylation of the neuronal tau protein and its interaction with the N-terminal domain of apolipoprotein E. *J. Biol. Chem.* **2010**, *285*, 33435–33444. [CrossRef] [PubMed]

47. Schwalbe, M.; Biernat, J.; Bibow, S.; Ozenne, V.; Jensen, M.R.; Kadavath, H.; Blackledge, M.; Mandelkow, E.; Zweckstetter, M. Phosphorylation of human Tau protein by microtubule affinity-regulating kinase 2. *Biochemistry* **2013**, *52*, 9068–9079. [CrossRef] [PubMed]

48. Qi, H.; Prabakaran, S.; Cantrelle, F.X.; Chambraud, B.; Gunawardena, J.; Lippens, G.; Landrieu, I. Characterization of neuronal Tau protein as a target of extracellular signal-regulated kinase. *J. Biol. Chem.* **2016**, *291*, 7742–7753. [CrossRef] [PubMed]

49. Lippens, G.; Landrieu, I.; Smet, C.; Huvent, I.; Gandhi, N.S.; Gigant, B.; Despres, C.; Qi, H.; Lopez, J. NMR meets Tau: Insights into its function and pathology. *Biomolecules* **2016**, *6*. [CrossRef] [PubMed]

50. Limatola, A.; Eichmann, C.; Jacob, R.S.; Ben-Nissan, G.; Sharon, M.; Binolfi, A.; Selenko, P. Time-resolved NMR analysis of proteolytic alpha-synuclein processing in vitro and in cellulo. *Proteomics* **2018**, *18*, e1800056. [CrossRef]

51. Barghorn, S.; Biernat, J.; Mandelkow, E. Purification of recombinant tau protein and preparation of Alzheimer-paired helical filaments in vitro. *Methods Mol. Biol.* **2005**, *299*, 35–51. [PubMed]

52. Timm, T.; Li, X.Y.; Biernat, J.; Jiao, J.; Mandelkow, E.; Vandekerckhove, J.; Mandelkow, E.M. MARKK, a Ste20-like kinase, activates the polarity-inducing kinase MARK/PAR-1. *EMBO J.* **2003**, *22*, 5090–5101. [CrossRef] [PubMed]

53. Eliezer, D.; Barre, P.; Kobaslija, M.; Chan, D.; Li, X.; Heend, L. Residual structure in the repeat domain of tau: Echoes of microtubule binding and paired helical filament formation. *Biochemistry* **2005**, *44*, 1026–1036. [CrossRef] [PubMed]

54. Schneider, C.A.; Rasband, W.S.; Eliceiri, K.W. NIH Image to ImageJ: 25 years of image analysis. *Nat. Methods* **2012**, *9*, 671–675. [CrossRef] [PubMed]

55. Delaglio, F.; Grzesiek, S.; Vuister, G.W.; Zhu, G.; Pfeifer, J.; Bax, A. NMRPipe: A multidimensional spectral processing system based on UNIX pipes. *J. Biomol. NMR* **1995**, *6*, 277–293. [CrossRef] [PubMed]

56. Johnson, B.A. Using NMRView to visualize and analyze the NMR spectra of macromolecules. *Methods Mol. Biol.* **2004**, *278*, 313–352. [CrossRef] [PubMed]

International Journal of
Molecular Sciences

MDPI

Review

The Inescapable Effects of Ribosomes on In-Cell NMR Spectroscopy and the Implications for Regulation of Biological Activity

David S. Burz, Leonard Breindel and Alexander Shekhtman *

Department of Chemistry, University at Albany, State University of New York, 1400 Washington Ave., Albany, NY 12222, USA; dsburz@albany.edu (D.S.B.); lbreindel@albany.edu (L.B.)
* Correspondence: ashekhtman@albany.edu

Received: 5 February 2019; Accepted: 9 March 2019; Published: 14 March 2019

Abstract: The effects of RNA on in-cell NMR spectroscopy and ribosomes on the kinetic activity of several metabolic enzymes are reviewed. Quinary interactions between labelled target proteins and RNA broaden in-cell NMR spectra yielding apparent megadalton molecular weights in-cell. The in-cell spectra can be resolved by using cross relaxation-induced polarization transfer (CRINEPT), heteronuclear multiple quantum coherence (HMQC), transverse relaxation-optimized, NMR spectroscopy (TROSY). The effect is reproduced in vitro by using reconstituted total cellular RNA and purified ribosome preparations. Furthermore, ribosomal binding antibiotics alter protein quinary structure through protein-ribosome and protein-mRNA-ribosome interactions. The quinary interactions of Adenylate kinase, Thymidylate synthase and Dihydrofolate reductase alter kinetic properties of the enzymes. The results demonstrate that ribosomes may specifically contribute to the regulation of biological activity.

Keywords: Ribosome; mRNA; rRNA; Thioredoxin; Adenylate kinase; Thymidylate synthase; Dihydrofolate reductase; cross-correlated relaxation; protein interactions; protein structure-function; enzyme activity; enzyme kinetics; NMR spectroscopy

1. Introduction

For the past two decades, in-cell NMR spectroscopy has been used to investigate the structure, dynamics and interaction surfaces of proteins inside living cells [1–7]. In recent years a few intrinsically disordered proteins, IDPs, such as alpha-synuclein [5], Pup [8], and FG repeats [9,10] and folded proteins, such as GB1 [11] and SOD1 [12], have provided in-cell NMR spectra of satisfactory quality for quantitative analysis. However, the in-cell NMR spectra of most folded proteins are poorly resolved when employing the pulse sequences typically used to study proteins in vitro [13]. Binding interactions between the target protein and intracellular constituents result in macromolecular complexes with apparent molecular weights on the order of 1 MDa [14,15] that scale linearly with intracellular viscosity and are consistent with in vitro apparent molecular weights of 300–400 kDa [16,17]. As larger species tumble more slowly the result is a widespread broadening of in-cell NMR spectral peaks [14,16]. These specific low-affinity interactions, dubbed quinary interactions, are omnipresent due to the high concentration of interacting species, which provide the chemical energy for binding interactions [18–20].

To be detectable by in-cell NMR, target proteins have to be present in-cell at concentrations $\geq$10 μM [13,21–23]. What intracellular species exist at sufficiently high concentrations to give rise to protein quinary structures? Genomic DNA is too large (>10 MDa), has too low an abundance and is largely inaccessible in eukaryotic cells [24]. Proteins, with an average molecular mass of ~50 kDa [25], and tRNAs, ~20 kDa [24], will not form complexes of the size observed. That leaves mRNA, 100–500 kDa, and rRNA, up to 5 MDa, as the most likely candidates for the interacting complement to protein quinary structural complexes.

The intracellular concentrations of mRNA have been estimated to range from 2–20 μM in prokaryotes and 50–500 nM in eukaryotes [24]. Ribosome concentrations in prokaryotes and eukaryotes can exceed 10 μM and 1 μM, respectively [25]. These concentrations are high enough to ensure a wide range of binding interactions with target proteins that are introduced into or over-expressed in cells. The ubiquity of these interactions forms the bedrock for quinary structural states that represent the primary conformation adopted by most proteins in cells [26].

Over the past few years, work in our laboratory has suggested that RNA, in particular ribosomes, plays a major role in establishing protein quinary structures [14,26]. This conclusion is in general agreement with mass spectroscopic studies of mRNA- and ribo-interactomes [27–30] in which hundreds of eukaryotic proteins bound to either mRNA or ribosomes were identified and did not possess obvious RNA binding motifs. Such observations have provided a glimpse of insight into the physical complexity of quinary interactions [31–34]. Additional evidence suggests that the RNA-bound quinary state may have a different activity than the unbound state of the protein studied in vitro [26,35]. In this article, we will review the evidence for implicating RNA as an integral component that interacts with folded proteins to establish quinary structure (Table 1) and show that the biological activity of a protein is altered when bound to ribosomes.

Table 1. Summary of protein quinary interactions.

Protein	Binds	Effect of RNA-Binding
Ubiquitin (Ubq)	Total RNA [14,35] mRNA [28]	Blocks polyubiquitination sites, increases apparent MW [14]
Thioredoxin (Trx)	Total RNA [14], mRNA [28]	Increases apparent MW [14] Antibiotic binding to ribosome alters quinary structure [36]
Adenylate kinase (ADK)	Total RNA [14] mRNA [28] Ribosome [26,29]	Increases apparent MW [14] Noncompetitive kinetic inhibitor [26]
Dihydrofolate reductase (DHFR)	mRNA [37,38] Ribosome [26]	Competitive kinetic inhibitor [26]
Thymidylate synthase (TS)	Total RNA [26] mRNA [28,39] Ribosome [26,29]	Uncompetitive kinetic activator [26]

2. Protein-RNA Interactions Broaden Target Protein NMR Spectra

Bertrand et al. [40] noted that changing the carbon source during growth of the yeast *Pichia pastoris*, *P. pastoris*, altered the intracellular distribution of the uniformly labeled overexpressed target protein Ubiquitin, [*U*- ^{15}N] Ubq, from *Saccharomyces cerevisiae*, *S. cerevisiae*. The in-cell ^{1}H-^{15}N heteronuclear single quantum coherence, HSQC, the spectrum of Ubq acquired from cells grown in methanol displays many broadened and missing peaks suggesting that Ubiquitin interacts with large intracellular complexes (Figure 1A). The dispersion of the detectable peaks indicates that Ubiquitin is well folded, but background signals from small ^{15}N labeled metabolites, which dominate the central region of the spectra, impede high-resolution analysis. For cells grown on methanol and dextrose, the spectrum is undetectable. The in-cell NMR spectrum of Ubq collected 48 h post-induction (Figure 1B) contains stronger signals suggesting that a larger fraction of the population is free to tumble inside the cells. By overexpressing Ubq for a very long period most of the binding sites become saturated allowing free Ubiquitin to be observed.

To determine if these results are due to Ubq–RNA quinary interactions, in vitro ^{1}H-^{15}N HSQC NMR spectra were collected on [*U*- ^{15}N] Ubq in the absence and presence of total RNA prepared from yeast cells grown in buffered methanol medium, RNA$_{BMM}$, and in buffered methanol/dextrose medium, RNA$_{BMDM}$ [35]. The effect of RNA on the HSQC NMR spectra was dramatic. Consistent with in-cell observations [40], in the presence of 30 mg/mL of RNA$_{BMM}$ a subset of Ubq crosspeaks were broadened (Figure 1C) suggesting a specific interaction between the labeled target and RNA [35]. In the presence of 30 mg/mL of RNA$_{BMDM}$ all of the spectral peaks disappeared (Figure 1D).

There were conspicuous differences between the two RNA preparations: RNA$_{BMM}$ contained preprocessed large ribosomal and mRNA that was absent from RNA$_{BMDM}$ (Figure 1E). Control HSQC spectra collected in the presence of up to 50 mg/mL of chondroitin sulfate, a glycosylate linear polyanion, did not affect the basis spectrum suggesting that the Ubq-RNA interaction is specific [35]. The conclusion was that Ubq quinary interactions were regulated by the total cellular RNA content, which was, in turn, regulated by the growth conditions, and that the affinity of the interaction increased in the presence of fully processed RNA. The use of total RNA preparations successfully recapitulated in-cell observations and provided an in vitro platform for further investigating the role of RNA in promoting and maintaining quinary structural states.

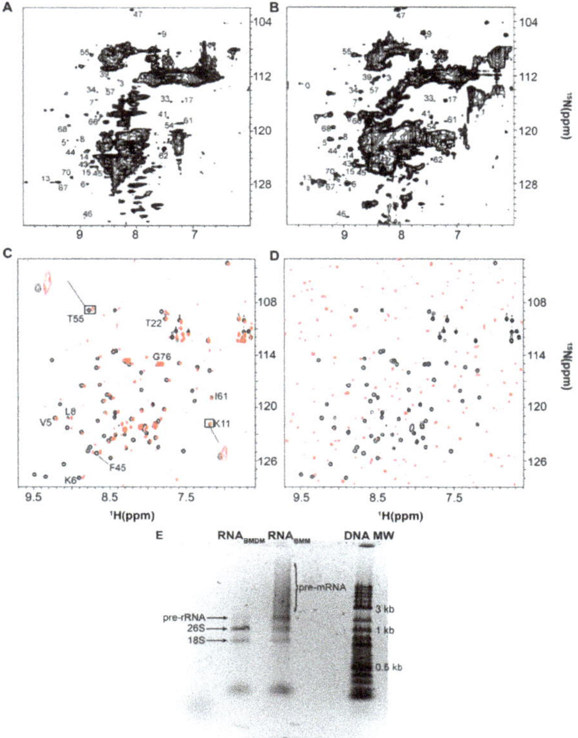

Figure 1. Total cellular RNA alters in vitro spectra of Ubiquitin, Ubq. (**A**) In-cell ^{1}H-^{15}N heteronuclear single quantum coherence, HSQC, NMR spectra of [U-^{15}N] Ubq in *P. pastoris* after 24 h of methanol induction and (**B**) 48 h of methanol induction. (**C**) Overlay of the in vitro ^{1}H-^{15}N HSQC spectra of 10 μM [U-^{15}N] Ubq in the absence (black) and presence (red) of 30 mg/mL of RNA$_{BMM}$ and (**D**) 30 mg/mL of RNA$_{BMDM}$. Insets in panel C show a broadening of selected residues of free Ubq (black) due to the interaction with RNA$_{BMM}$ (red). (**E**) RNA from yeast cells grown with methanol, RNA$_{BMM}$, contains an amount of pre-mRNA and pre-rRNA larger than that of RNA from cells grown with a methanol/dextrose carbon source, RNA$_{BMDM}$. DNA MW indicates molecular weight markers. The numbers in panels A, B and C indicate some of the peak assignments. Panels A and B are adapted from Bertrand et al. (2012) [40]. Panels **C**, **D** and **E** are adapted from Majumder et al. (2016) [35].

3. Resolving Target Protein Bound to RNA

The problem of widespread target protein in-cell HSQC NMR signal broadening is not limited to yeast. Indeed, virtually all proteins display these spectral characteristics in both mammalian and bacterial cells [14,41,42]. The absence of widespread line broadening in early experiments performed in *E. coli* was due to the fact that the overexpressed proteins leaked out of the cells during in-cell NMR experiments [43] or overexpression of labeled target exceeded 100 μM [13], which is $\geq$10 times greater than the estimated dissociation constant of 1–10 μM for target protein quinary interactions. At this concentration, in-cell NMR signal intensity is enhanced by a population of the unbound protein resulting in a greater number of sharper spectral peaks. At lower intracellular concentrations binding of the labeled target is stoichiometric. Due to the high concentration of RNA present in cells, line broadening is inevitable for proteins expressed at physiological levels. To ascribe biological relevance to the structural interactions revealed by in-cell NMR spectra it was necessary to adopt methods for detecting large labeled targets at or near physiological concentrations.

Peak broadening is due to the formation of massive quinary interaction complexes. The large MW species tumble more slowly and exhibit a reduced transverse relaxation time for the NMR signal, T2 [44,45]. T2 depends on the rotational diffusion of a molecule in solution and is inversely related to the rotational correlation time, τ_c [45]. Shorter T2 values cause the NMR signal from larger molecules to decay more rapidly and lead to extensive line broadening [44]. This effect is pronounced in the case of folded proteins where all nuclei experience global rotation. Notable exceptions include intrinsically disordered proteins, IDPs, and protein with intrinsically disordered regions, IDRs [46]. These proteins lack persistent secondary or higher structure, possess fast local dynamics, and fail to interact with intracellular constituents resulting in in-cell spectra that are much sharper than those typically observed for folded proteins [47].

HSQC and heteronuclear multiple quantum coherence, HMQC, pulse sequences [45], originally used for in-cell NMR spectroscopy [48], use insensitive nuclei enhanced by polarization transfer, INEPT, pulse sequences to transfer magnetization from protons to heteronuclei, but the efficiency of INEPT deteriorates with decreasing T2 [49]. Transverse relaxation-optimized spectroscopy, TROSY, which suppresses transverse nuclear spin relaxation in heteronuclear NMR experiments during evolution and acquisition cycles [50] in combination with ^{15}N-edited cross relaxation-induced polarization transfer, CRINEPT, NMR spectroscopy [51,52], which increases the efficiency of magnetization transfers between heteronuclei, can be used to improve the resolution and sensitivity of in-cell NMR experiments for large complexes. Further improvement in sensitivity can be achieved by optimizing the CRINEPT-like magnetization transfer delay time in the ^{1}H-^{15}N CRINEPT-HMQC-TROSY pulse sequence, and by employing REDuced PROton density (REDPRO) labeling [53], which exchanges alpha and beta protons of amino acids for deuterons to minimize proton relaxation. The resulting in-cell ^{1}H-^{15}N CRINEPT-HMQC-TROSY pulse sequence when applied to [U-^{2}H, ^{15}N] labeled target protein yields a spectrum in which most of the target protein crosspeaks are resolved.

The improvement in spectral resolution is shown in Figure 2 for *Escherichia coli*, *E. coli*, Adenylate kinase, ADK. Using the ^{1}H-^{15}N HSQC pulse sequence the in vitro ^{1}H-^{15}N correlation spectrum is well-resolved (Figure 2A) but cannot be observed in *E. coli* cells (Figure 2B). Majumder et al. 2015 [14] utilized ^{1}H-^{15}N CRINEPT–HMQC–TROSY NMR to investigate uniformly ^{2}H and ^{15}N labeled, [U-^{2}H, ^{15}N], ADK in *E. coli* and was able to resolve many of the target protein peaks (Figure 2C). Similar results were obtained by using ^{1}H-^{15}N CRINEPT-HMQC-TROSY NMR to examine bacterial Thioredoxin, Trx and FK506 binding protein, FKBP, in *E. coli*, and human Ubq in HeLa cells [14]. Most importantly, the ^{1}H-^{15}N CRINEPT–HMQC–TROSY NMR spectrum of 10 μM [U-^{2}H, ^{15}N] ADK collected in vitro in the presence of 2.5 μM ribosomes exhibited broadened peaks that largely coincide with the in-cell spectrum (Figure 2D). This observation supports the idea that RNA, specifically ribosomes in the case of ADK, are the binding complement that gives rise to quinary interactions.

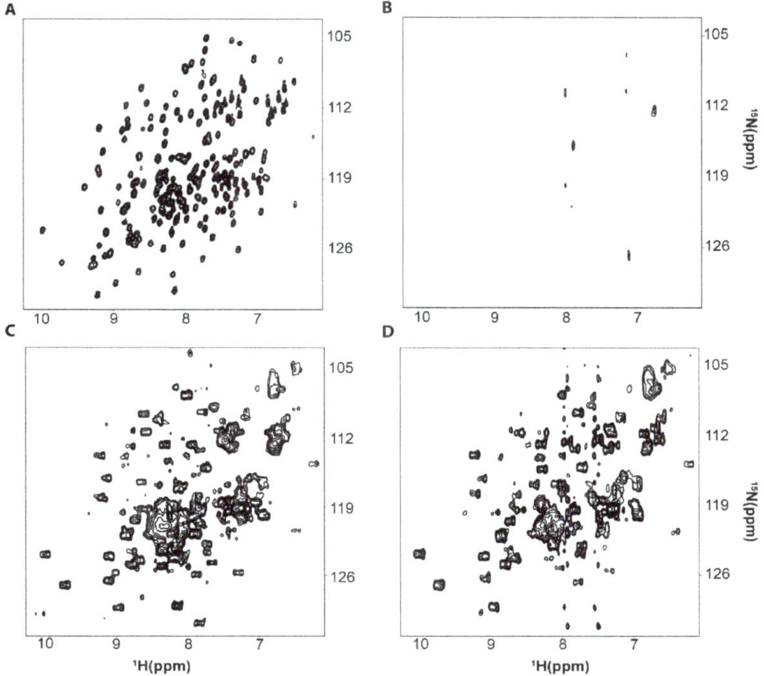

Figure 2. ^{1}H-^{15}N CRINEPT-HMQC-TROSY improves in-cell NMR spectral resolution. (**A**) Lysate HSQC spectrum of [*U*-^{15}N] Adenylate kinase, ADK. (**B**) In-cell ^{1}H-^{15}N HSQC spectrum of [*U*-^{15}N] ADK overexpressed for 16–18 h. (**C**) In-cell ^{1}H-^{15}N CRINEPT-HMQC-TROSY spectrum of [*U*-^{2}H, ^{15}N] ADK overexpressed for 16–18 h. (**D**) In vitro ^{1}H-^{15}N CRINEPT-HMQC-TROSY spectrum of 10 μM purified [*U*-^{2}H, ^{15}N] ADK in the presence of 2.5 μM ribosomes. The peak shapes in **C** and **D** arise from a population of free and bound species due to the high concentration of target protein (>100 μM).

4. Target Protein-RNA Complexes Exhibit Megadalton Apparent Molecular Masses

Optimizing the CRINEPT transfer delay time, T_{opt}, can provide an estimate of the apparent molecular weight of the target protein. Theoretically [52,54] T_{opt} is a solution of

$$R_c\left[sinh\left(2R_cT_{opt}\right)\right] + \pi J_{NH}\left[sin\left(2\pi J_{NH}T_{opt}\right)\right] = 2R_H\left[sinh^2\left(R_cT_{opt}\right) + sin^2\left(\pi J_{NH}T_{opt}\right)\right] \tag{1}$$

where R_c is the relaxation rate resulting from the cross-correlation between ^{15}N–^{1}H dipole–dipole coupling and amide proton chemical shift anisotropy, R_H is the transverse relaxation rate of the amide protons and J_{NH} is a scalar ^{15}N–^{1}H coupling constant. R_c and R_H are related to the rotational correlation time, τ_c, by $R_c = 1.7\tau_c B_0$ and $R_H = \tau_c(0.8B_0^2 + 1.7)$, where τ_c is in nanoseconds, B_0 is the strength of the magnetic field in gigahertz and R_c and R_H are in seconds. In combination with the Debye–Stokes–Einstein relation [55]

$$\tau_c = \left(4\pi\eta R_H^3\right)/3kT. \tag{2}$$

where η is the viscosity of the medium, R_H is the hydrated radius of the protein, k is the Boltzman constant and T is absolute temperature, the apparent molecular weight, MW_{app}, of protein inside cells, bound to RNA, or in viscous glycerol solutions can be estimated. Solving T_{opt} for a range of τ_c values will yield Stokes radii that can be used to approximate MW_{app} by assuming a generic value for the partial specific volume of a protein equal to 0.73 cc/g [51].

Data showing the dependence of T_{opt} on the MW_{app} of *E. coli* Trx, measured in vitro with an increasing amount of glycerol, which restricts the rate of tumbling, is shown in Figure 3A [14]. The experimental data agree well with the theoretical curve generated using Equation (1). T_{opt} was measured for *E. coli* ADK, MW 23.5 kDa, and found to be $\leq$1.2 ms (Figure 3B); the lack of a maximum in the in-cell *E. coli* buildup curve implies that the apparent molecular weight is $\geq$1.2 MDa. Transfer times shorter than 1.2 ms interfere with CRINEPT pulses and limit the ability to collect data. In vitro in the presence of total *E. coli* RNA T_{opt} was 2.5 ms (Figure 3C), which corresponds to an apparent molecular weight, MW_{app}, of ~0.4 MDa. Because ADK was present in molar excess over total RNA, the resolved MW_{app} reflects a population of free and RNA-bound ADK. Correcting for an intracellular viscosity of 3–4 cP, yields an in-cell MW_{app} of ~1.4 MDa. *E. coli* Trx, MW 11.8 kDa, exhibited a T_{opt} of 1.3 ms in-cell, which corresponded to an MW_{app} of ~1.1 MDa, and in the presence of total *E. coli* RNA the uncorrected MW_{app} was ~0.3 kDa (Figure 3A), which translates to an in-cell MW_{app} of ~1.1 MDa. Given that the approximate MW of an *E. coli* ribosome is 1.3 MDa [25], MW_{app} for the target proteins observed in-cell and in vitro in the presence of total RNA are consistent with the formation of protein-ribosomal complexes.

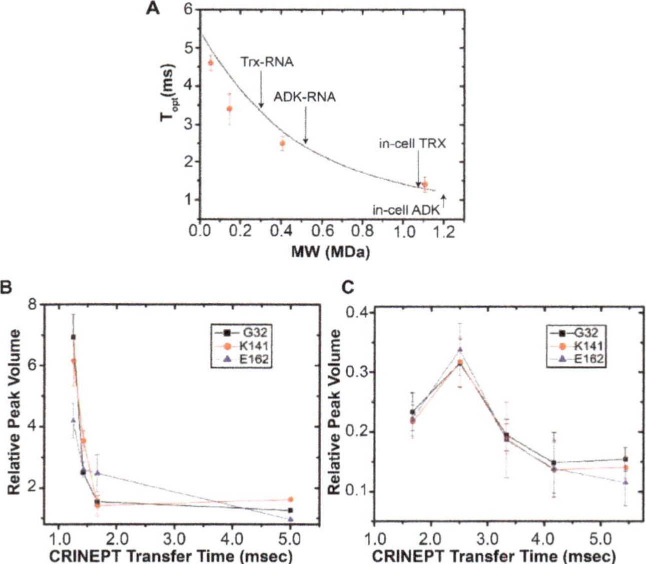

Figure 3. Optimizing the CRINEPT transfer delay time yields in-cell target protein apparent molecular weights. (**A**) The dependence of T_{opt} on the apparent in-cell molecular weight, MW_{app} at 700 MHz. T_{opt} was experimentally determined at 5 °C (red symbols) by using 100 µM [U-^{2}H, ^{15}N] Trx dissolved in 10 mM potassium phosphate buffer, pH 6.5, containing 30, 65, 75, and 85% (*w/w*) d$_5$-glycerol with corresponding viscosities of 4, 34, 92, and 343 cP, respectively [56]. The MW_{app} of *E. coli* ADK and Trx in-cell and in vitro in the presence of total *E. coli* RNA, uncorrected for intracellular viscosity, are indicated. (**B,C**) The relative volumes of the G32, K141 and E162 peaks in the ^{1}H-^{15}N CRINEPT-HMQC–TROSY spectra of [U-^{2}H, ^{15}N] ADK collected in-cell (**B**) and in vitro at 20 µM in the presence of 50 µg of total RNA (**C**) are plotted against CRINEPT transfer delay times. In (**B**) an in-cell value of 1.2 ms was assigned because shorter transfer delay times interfere with CRINEPT pulses and limit the ability to acquire data. An endogenous tryptophan indole amide peak in the in-cell spectra was used as a reference. Panels (**A–C**) are adapted from Majumder et al. (2015) [14].

5. RNA-Mediated Quinary Interaction Surfaces

5.1. Adenylate Kinase

The interacting surfaces of a target protein in-cell can be determined by using STructural INTeraction, STINT, NMR [4,57–59], which quantitates the changes in individual crosspeaks between the free and bound conformations. The signal from target protein surface residues is altered when engaged in binding interactions. For quinary interactions, the changes in chemical shift and/or intensity between in vitro or lysate target protein crosspeaks are compared to those observed in-cell to identify the quinary interaction surface. Further changes in those surfaces in response to stimuli can be analyzed by using singular value decomposition, SVD, which distinguishes concentration-dependent from concentration-independent changes in crosspeaks over time as the concentration of the stimulus increases [58,60]. STINT-NMR was used to investigate the quinary structure of ADK and Trx, and the changes in quinary structure in response to ribosomal-binding antibiotics [14,36].

ADK catalyzes the transfer of a phosphate from ATP to AMP to create two ADP molecules [61]. In the absence of bound substrate, the enzyme exists in an open conformation in which the ATP and AMP binding domains are maximally separated; substrate binding reorients the domains closer together resulting in a closed conformation [62,63]. The ^{1}H-^{15}N CRINEPT-HMQC-TROSY spectrum of *E. coli* ADK collected in *E. coli* cells [14] indicates an open conformation in agreement with in vitro observations. The spectral broadening was characteristic of intermediate exchange, implying an interaction dissociation constant between 1–10 µM. The chemical shifts of residues involved in domain closure were unchanged showing that macromolecular crowding does not perturb the tertiary structure of the enzyme. A subset of peak intensities was broadened in the in-cell spectrum of ADK relative to what was observed in vitro or in lysates (Figure 4A). The residues that undergo the most dramatic changes in intensities in-cell define the quinary interaction surface (Figure 4B). This surface lies proximal to the AMP binding region in the CORE domain leaving the active sites of ADK unaffected and free to bind ATP and AMP.

The exact nature of the interaction was further clarified in vitro by using NMR (Figure 5A) and fluorescence titration (Figure 5B) to measure the binding of ADK to ribosomes. ADK was found to bind to ribosomes with a K_d of 3.7 ± 0.4 µM. The quinary contact surface identified by in-cell NMR therefore likely represents the ADK-ribosomal interface. To further interrogate the relationship between ADK and ribosomes, chloramphenicol, which binds to the large ribosomal subunit and increases the intracellular concentration of ATP [64], was introduced into *E. coli* and the resulting spectral changes in the [U-^{2}H, ^{15}N] ADK spectrum were analyzed.

The addition of chloramphenicol perturbed the cellular equilibrium between ATP, ADP and AMP and dramatically altered the in-cell NMR spectrum of ADK. Changes in chemical shifts consistent with ATP- and AMP-bound ADK were observed (Figure 5C). The spectrum was similar to that observed in vitro with 3 mM ATP and 200 µM AMP, and consistent with a closed conformation of ADK. Collectively the results suggest that ribosomes may regulate the activity of ADK directly through quinary interactions, which may alter the affinity of the enzyme for ATP, or indirectly by altering the concentration of free ATP available for binding.

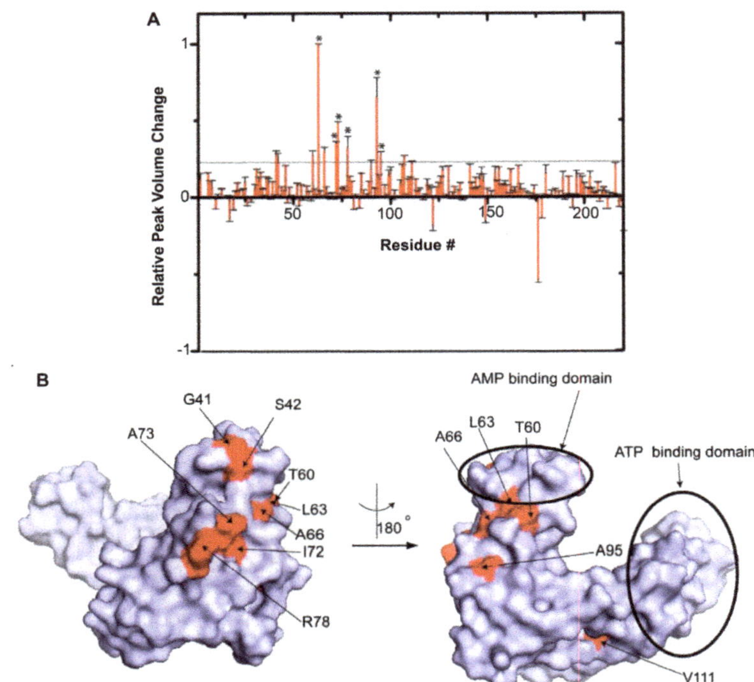

Figure 4. ADK quinary interaction surface does not block the active sites. (**A**) Relative changes in in-cell ^{1}H-^{15}N CRINEPT–HMQC–TROSY peak intensities of [*U*- ^{2}H, ^{15}N] ADK residues due to ribosome-mediated quinary interactions. The threshold line delineates residues whose NMR peaks undergo significant broadening. Residues that are affected by the interaction of ADK with total RNA are indicated with asterisks. (**B**) Residues involved in quinary interactions (red), mapped onto the molecular surface of ADK (Protein Data Bank, PDB, entry 4AKE), lie in the CORE domain of ADK. Panels (**A**,**B**) are adapted from Majumder et al. (2015) [14].

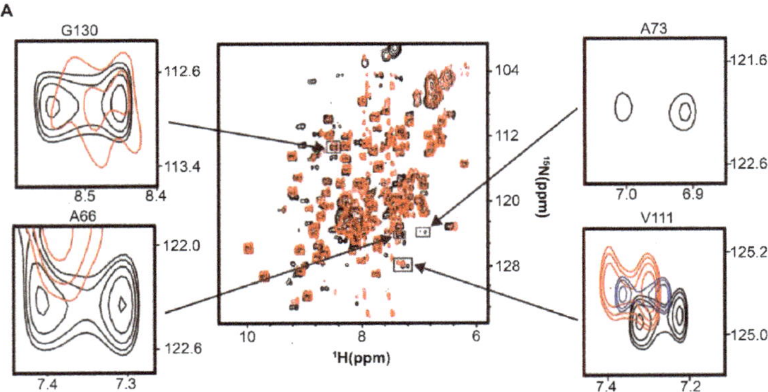

Figure 5. *Cont.*

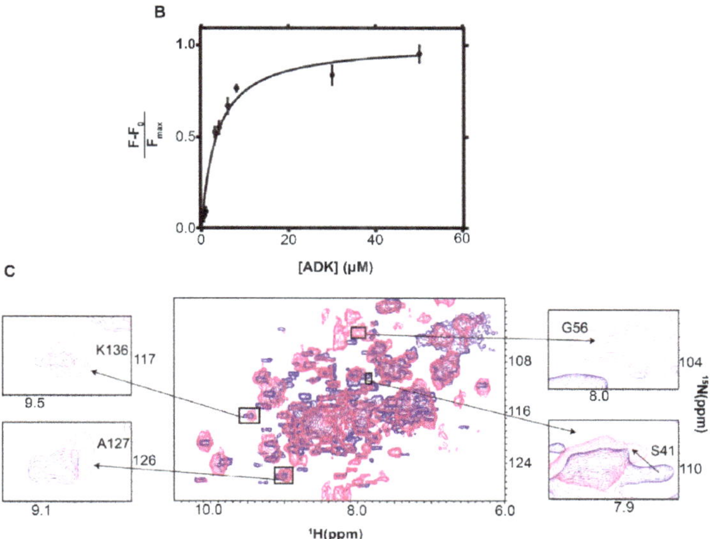

Figure 5. Quinary interactions of ADK in *E. coli*. (**A**) (Center) Overlay of in vitro ^{1}H-^{15}N CRINEPT-HMQC-TROSY spectra of 10 μM [*U*- ^{2}H, ^{15}N] ADK without (black) and with 2.5 μM ribosome (red). Surrounding panels show overlays of individual residues including in-cell NMR peaks (blue). (**B**) Fluorescence titration of 0.5 μM ribosome with ADK. Tryptophan fluorescence was measured at an emission wavelength of 350 nm by using an excitation wavelength of 280 nm. Curve fitting to a single site-binding isotherm yielded a K_d of 3.7 ± 0.4 μM. F_o is the fluorescence in the absence of ADK, and F_{max} is the maximum fluorescence of the ADK−ribosome complex. Fluorescence titration experiments were performed in triplicate. (**C**) Overlay of the in-cell ^{1}H-^{15}N CRINEPT-HMQC-TROSY spectra of [*U*- ^{2}H, ^{15}N] ADK in the absence (blue) and presence (magenta) of 100 μg/mL chloramphenicol. K136 and A127 (left insets) in chloramphenicol treated cells exhibit chemical shift changes consistent with ATP bound ADK; G56 and S41 peaks (right insets) exhibit chemical shift changes consistent with AMP bound ADK. Panels A and B are adapted from DeMott et al. (2017) [26]. Panel (**C**) is adapted from Majumder et al. (2015) [14].

5.2. Thioredoxin

E. coli Trx, is a 12 kDa protein with redox activity that maintains a reducing environment inside the cell by means of active site cysteines. The ^{1}H-^{15}N CRINEPT-HMQC-TROSY spectrum of [*U*- ^{2}H, ^{15}N] *E. coli* Trx collected in *E. coli* cells [14] exhibited broad peaks at positions close to those observed in cell lysates (Figure 6A). The in-cell concentration of Trx was ~300 μM. Some of the in-cell NMR resonances exhibit two maxima, corresponding to fast and slow transverse relaxing components of crosspeaks (Figure 6B), suggesting that free cytosolic Trx is in exchange with a complex inside the cells. Despite this heterogeneity, only a subset of residues was broadened (Figure 6C). The results indicate that the quinary interaction surface of the molecule overlaps with the CGPC motif active site and adjacent regions (Figure 6D).

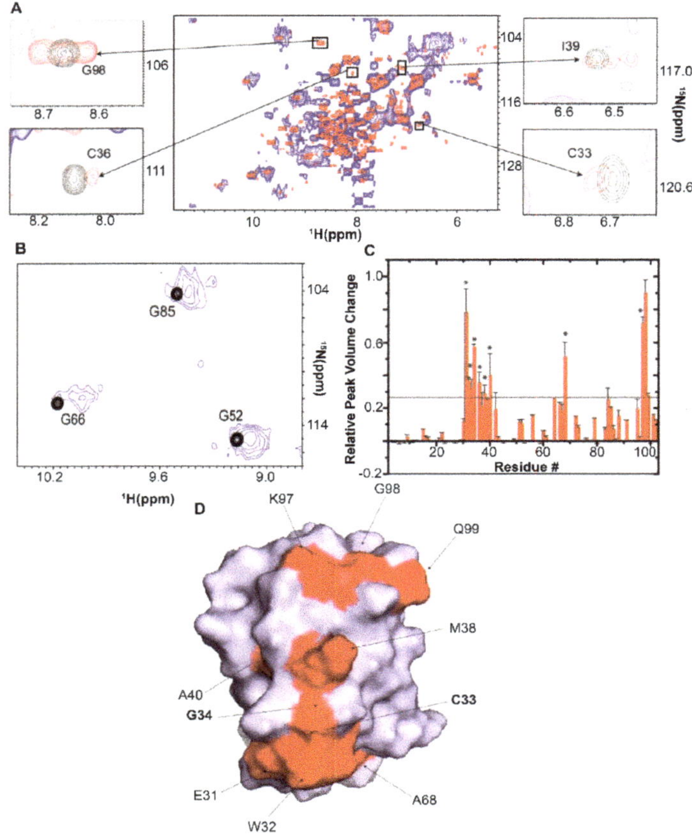

Figure 6. Quinary interactions of Trx in *E. coli*. (**A**) Overlay of the in-cell ^{1}H-^{15}N CRINEPT–HMQC–TROSY spectra of [U-^{2}H, ^{15}N] Trx (blue) and that of the cellular lysate (red). The insets show overlays of the boxed regions of the in-cell spectrum (blue) and the corresponding regions of the ^{1}H-^{15}N CRINEPT–HMQC–TROSY spectrum of lysate (red) and the ^{1}H-^{15}N HSQC spectrum of purified Trx in 10 mM potassium phosphate buffer (pH 6.5) (black). The intensities of the C33, C36, I39, and G98 peaks, residues involved in quinary interactions, are broadened in-cell. (**B**) Overlay of the ^{1}H-^{15}N CRINEPT–HMQC–TROSY spectrum of [U-^{2}H, ^{15}N] Trx in *E. coli* (blue) with crosspeaks from the ^{1}H-^{15}N HSQC spectrum of purified [U-^{2}H, ^{15}N] Trx in 10 mM potassium phosphate buffer, pH 6.5 (black). G52, G66, and G85 exhibit broad in-cell peaks characteristic of multiple conformations of Trx in fast exchange on the NMR time scale, implying that the quinary interactions are inherently transient and dynamic. (**C**) Relative changes in in-cell ^{1}H-^{15}N CRINEPT–HMQC–TROSY crosspeak intensities of [U-^{2}H, ^{15}N] Trx residues due to quinary interactions. The horizontal threshold differentiates residues whose NMR peaks undergo significant broadening. Residues annotated with asterisks are also affected in total RNA-bound Trx. (**D**) Residues involved in the quinary interactions (red) are mapped onto the molecular surface of Trx (PDB entry 1X0B); active site residues, C33 and G34, are in bold. The figure is adapted from Majumder et al. (2015) [14].

To determine if RNA is a component of Trx quinary interactions, ^{1}H-^{15}N CRINEPT-HMQC-TROSY spectra of 15 µM [U-^{2}H, ^{15}N] Trx were collected in the presence of 30 mg/mL of both *E. coli* and *S. cerevisiae* total RNA [14]. The indole NH of W29 exhibited the same downfield shift in the in vitro RNA-bound and in-cell NMR spectra, while the indole NH of W32 and backbone amide peaks of E31,

C33, C36, K37, I39, and A40 were broadened in a manner similar to the quinary interaction observed in-cell. Treating a mixture of purified Trx and total *E. coli* RNA with RNase A yielded a pool of small RNAs and nucleotides. If RNA oligonucleotides act as ligands, RNase treatment would increase the fraction of RNA-bound Trx due to the increase in the molar concentration of total RNA. Indeed, changes in the ^{1}H-^{15}N CRINEPT-HMQC-TROSY spectrum indicated an increase in the population of bound Trx and a reduced, ~20 kDa, MW_{app} both of which are expected to result from oligonucleotide binding. Total RNA and RNase-treated total RNA perturbed the same subset of peaks, indicating a specific quinary interaction surface for Trx.

To identify the RNA complement to Trx quinary interactions ^{1}H-^{15}N CRINEPT-HMQC-TROSY spectra of 150 µM [*U*-^{2}H, ^{15}N] Trx were acquired in the absence and presence of 10 µM ribosomes [26]. No peak broadening was observed implying that there was no specific interaction between Trx and ribosomes. In studies of the mRNA interactome, the eukaryotic homologue of Trx was shown to bind to mRNA [27,65–67]. The conclusion was that the Trx-RNA interaction previously identified was likely mediated by mRNA.

The putative Trx-mRNA interactions provided an opportunity to test whether quinary structures can be indirectly affected through mRNA-ribosome interactions, specifically through the influence of ribosome binding antibiotics. Ribosome inhibition depends on how the antibiotic is bound: binding to the small, 30S, ribosomal subunit can affect mRNA-ribosomal interactions whereas binding to the large, 50S, subunit interferes with the peptidyl transferase activity [68]. In the absence of antibiotics it was expected that the quinary interactions of Trx would not vary over time, and because ribosome inhibitors can alter mRNA-ribosome interactions, Trx quinary interactions could be profoundly altered.

Using a bioreactor that monitors real-time changes in in-cell NMR spectra, Breindel et al. [36] administered tetracycline and streptomycin, which bind to the 30S subunit, and chloramphenicol, which binds to the 50S ribosomal subunit, to *E. coli* containing overexpressed [*U*-^{15}N] Trx. The in-cell ^{1}H-^{15}N CRINEPT-HMQC-TROSY spectra were analyzed by using SVD to identify concentration-dependent changes as the concentration of antibiotic increased. The spectra were extensively broadened in the presence of tetracycline (Figure 7A) and streptomycin (Figure 7B). SVD analysis showed a sharp drop in the Scree plot of singular values with poor linear fits, r^2 of 0.67 and 0.66 for tetracycline (Figure 7C) and streptomycin (Figure 7D) respectively, indicating specific changes in quinary interactions. Not unexpectedly, the addition of chloramphenicol, which does not disturb the binding of mRNA, resulted in a linear decrease in singular values, r^2 = 0.94, suggesting that Trx quinary interactions were not perturbed.

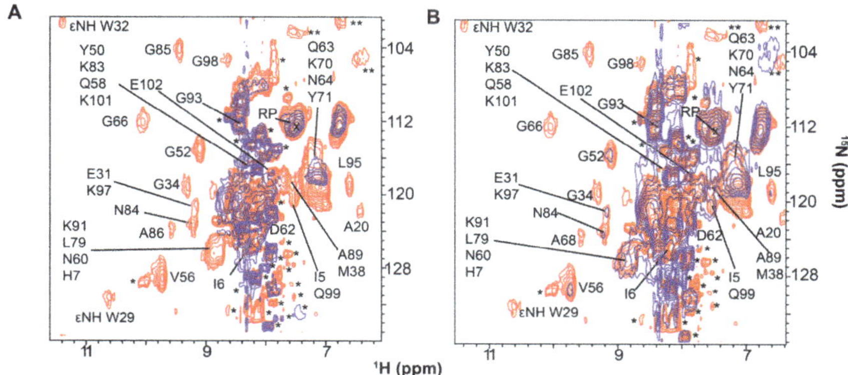

Figure 7. *Cont.*

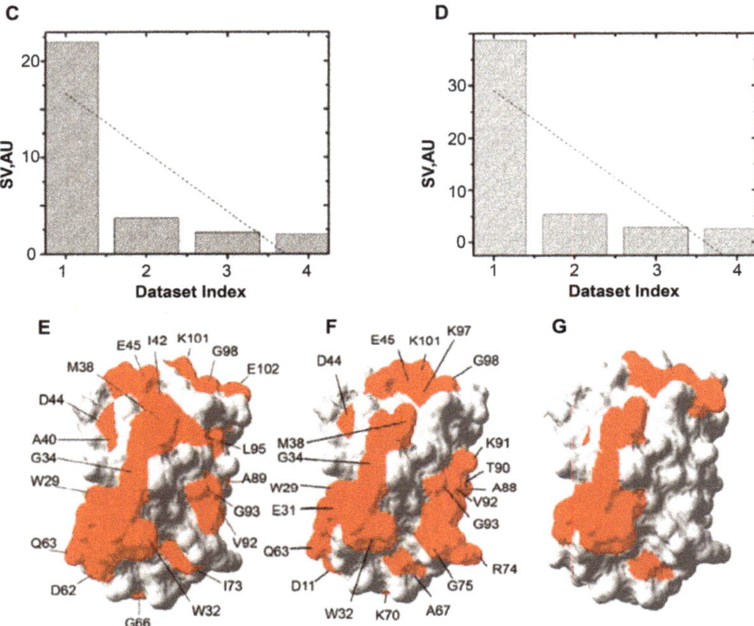

Figure 7. Binding of tetracycline and streptomycin to ribosomes changes the quinary structure of Trx in *E. coli*. (**A**) Overlay of the in-cell ^{1}H-^{15}N CRINEPT–HMQC–TROSY spectra of [U- ^{15}N] Trx without (red) and with (blue) tetracycline. (**B**) Overlay of the in-cell ^{1}H-^{15}N CRINEPT–HMQC–TROSY spectra of [U- ^{15}N] Trx without (red) and with (blue) streptomycin. Single and double asterisks indicate peaks from metabolites and unassigned side chain protons, respectively. The overlaid spectra are at the same contour levels. The reference peak used for peak intensity normalization is indicated by RP. (**C**) Distribution of singular values of each dataset index (binding mode) for Trx residues in the presence of tetracycline. (**D**) Distribution of singular values of each dataset index (binding mode) for Trx residues in the presence of streptomycin. (**E**) Residues involved in quinary interactions (red) due to the presence of tetracycline are mapped onto the molecular surface of Trx (Protein Data Bank entry 1X0B). (**F**) Residues involved in quinary interactions (red) due to the presence of streptomycin. (**G**) Quinary interaction surface (red) of Trx in the absence of antibiotics. Panels B–G are adapted from Breindel et al. (2017) [36].

Comparable changes in the Trx quinary interaction surface resulted from treating the cells with tetracycline and streptomycin respectively (Figure 7E,F). A large interaction surface containing negatively-charged and hydrophobic residues and a smaller patch containing positively-charged and hydrophobic residues are very similar to the Trx interaction surface in the absence of antibiotics (Figure 7G). A third adjoining surface, which does not participate in quinary interactions in the absence of antibiotics was differentially perturbed by tetracycline and streptomycin. In addition to surface residues, a number of buried residues underwent broadening in the presence of antibiotics, suggesting that some tertiary structural changes in Trx be occurring.

5.3. Dihydrofolate reductase and Thymidylate synthase

DeMott et al. 2017 [26] investigated two additional metabolic proteins: Dihydrofolate reductase [69,70], DHFR, and Thymidylate synthase [71], TS. The ^{1}H-^{15}N CRINEPT-HMQC-TROSY spectrum of DHFR was broadened in-cell and in vitro in the presence of ribosomes (Figure 8A). The ^{1}H-^{15}N HSQC in vitro NMR spectra of TS systematically broadened as the concentration of total

E. coli RNA was increased (Figure 8B,C). Ribosomes were subsequently shown to affect the kinetic activity of TS. Thus both enzymes acquired quinary structure by interacting with ribosomes.

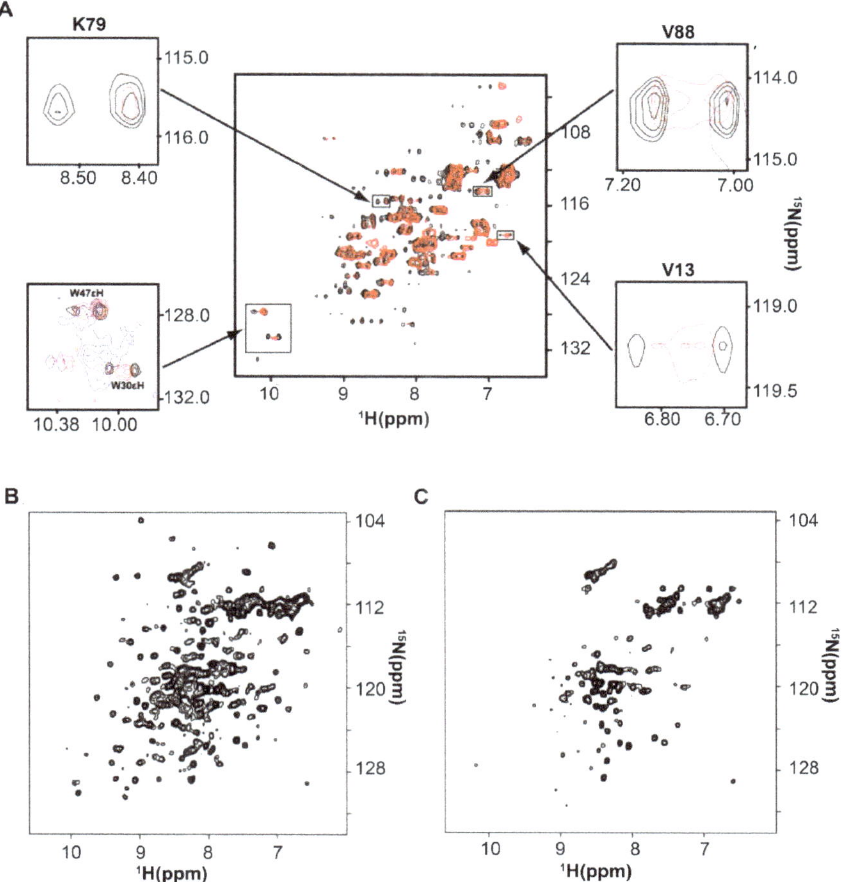

Figure 8. Dihydrofolate reductase, DHFR, and Thymidylate synthase, TS, engage in quinary interactions with RNA. (**A**) Overlay of *in vitro* ^{1}H-^{15}N CRINEPT–HMQC–TROSY spectra of 200 μM [*U*-^{2}H, ^{15}N] DHFR with 0.5 mM folate (black) and with 0.5 mM folate and 2.5 μ M ribosome (red). Insets show individual residue overlays that include in-cell NMR peaks (blue). Folate was added to increase the solubility of DHFR. (**B,C**) ^{1}H-^{15}N HSQC spectra of 50 μM [*U*- ^{15}N] TS with (**B**) 0 μg and (**C**) 135 μg of total *E. coli* RNA. The figure is adapted from DeMott et al. (2017) [26].

6. Ribosome-Mediated Regulation of Biological Activity

6.1. Adenylate Kinase

The studies delineated above showed that the quinary structures of ADK and Trx are mediated by protein-RNA interactions and that these structures can be affected directly and indirectly by perturbing the ribosome through the application of ribosomal-binding antibiotics. To investigate the possibility that the quinary state of the target protein may affect its biological activity, assays were performed in the presence of ribosome preparations.

DeMott et al., 2017 [26] found that in the absence of ribosomes the V_{max} for ADK reached a maximum at $\sim$1 mM and decreased at higher ATP concentrations, characteristic of noncompetitive substrate inhibition (Figure 9A). The kinetic profile suggests the presence of additional ATP binding sites [72,73]. Adding 1 μM ribosome decreased V_{max} by 50%, increased the substrate affinity by 30% and decreased the affinity of inhibitor binding, K_I, 6-fold (Table 2). The interaction between ADK and the ribosome does not occlude the active sites (Figure 4) but may preclude occupancy of allosteric binding sites. This would be consistent with a reduction in binding affinity for a second ATP binding site exemplified by K_I. Thus, the interaction between ADK and ribosomes establish a quinary activity state that reduces the V_{max} of ADK and mitigates substrate inhibition.

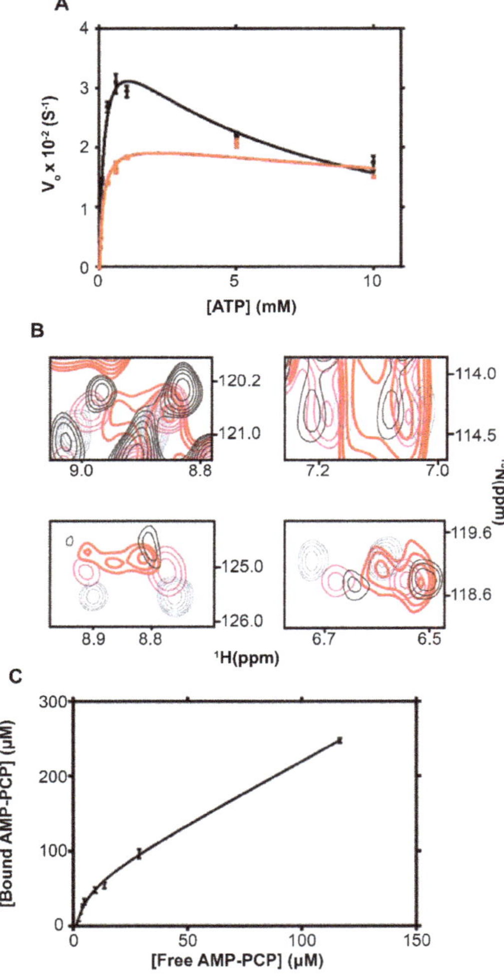

Figure 9. Ribosomes modulate ADK enzymatic activity. (**A**) Kinetic activity profile for ADK without (black) and with (red) 1 μM ribosome. (**B**) Overlays of in vitro ^{1}H-^{15}N CRINEPT–HMQC–TROSY spectra of 10 μM [*U*- ^{2}H, ^{15}N] ADK at 0 μM adenosine triphosphate, ATP, (blue), 20 μM ATP (magenta), 40 μM ATP (black), and 80 μM ATP plus 1 μM ribosome (red). (**C**) ATP analogue β,γ-methyleneadenosine 5′-triphosphate, AMP-PCP binding to ribosomes. The concentration of ribosomes was 2 μM. The figure is adapted from DeMott et al. (2017) [26].

Table 2. Kinetic Parameters Resolved for ADK, DHFR and TS in the Absence and Presence of Ribosomes.

Enzyme	[ribosome] (μM)	V_{max} (S^{-1})	$V_{max}^{Ribosome}/V^o_{max}$ [a]	K_M (μM) [b]	K_I (mM) [b]	R^2
ADK	0	$(4.2 \pm 0.2) \times 10^{-2}$	$\sim$0.5	180 ± 20	6.1 ± 0.8	0.98
	1	$(2.1 \pm 0.1) \times 10^{-2}$		130 ± 30	35 ± 2	0.96
DHFR	0	$(1.48 \pm 0.04) \times 10^{-4}$	$\sim$0.8	0.32 ± 0.04		0.95
	0.5	$(1.16 \pm 0.02) \times 10^{-4}$		3.5 ± 0.1		0.99
TS	0	$(9.7 \pm 0.4) \times 10^{-5}$	$\sim$20	5.4 ± 0.7		0.97
	0.5	$(2.0 \pm 0.2) \times 10^{-3}$		120 ± 20	$(3.9 \pm 0.5) \times 10^{-3}$	0.99

[a] $V_{max}^{Ribosome}$ and V_{max} are the maximum initial velocities with and without the ribosome. [b] Enzymatic parameters in the absence of ribosomes are consistent with those found at http://www.brenda-enzymes.org. Table adapted from DeMott et al. (2017) [26].

To assess the effect of ribosomes on the interaction between ATP and ADK, ^{1}H-^{15}N CRINEPT-HMQC-TROSY spectra of purified 10 μM [U-^{15}N] ADK were collected in the presence of increasing amounts of ATP. Systematic changes in the intensities and chemical shifts of interacting residues of ADK were observed as the concentration of ATP was increased from 0 to 40 μM. In the presence of 1 μM ribosome, the 80 μM ATP ^{1}H-^{15}N CRINEPT-HMQC-TROSY spectrum coincided more closely with the ADK spectrum acquired at 40 μ M ATP in the absence of ribosomes. The result suggests that the ribosome reduced the concentration of free ATP available for binding. This was consistent with the in-cell observation of an open conformation for ADK [14], which implied that only a small fraction of intracellular ATP binds to ADK, k_M = 51 μM [74], despite the fact that bacterial cells contain ~3 mM total ATP.

To investigate a possible mechanism for reducing the concentration of free ATP in *E. coli*, 2D ^{1}H-^{31}P-correlation NMR experiments were performed to quantify the binding of β,γ-methyleneadenosine 5'-triphosphate, AMP-PCP, a noncleavable ATP analogue, to ribosomes. The pH of the solution remained constant during the titration. Below 10 μM, the binding of AMP-PCP was fit to a single class of sites with an apparent K_d of 6 ± 2 μM; the inability to saturate the binding curve at higher concentrations prevented the estimation of an affinity constant for the weaker class of binding (Figure 9C). Thus it appears that ribosomes attenuate the in-cell activity of ADK by binding large amounts of ATP, thereby reducing the intracellular concentration of free ATP available to drive binding reactions, and suppress substrate inhibition through quinary interactions that reduce the affinity of regulatory sites.

6.2. Dihydrofolate Reductase and Thymidylate Synthase

TS and DHFR are functionally linked in the *de novo* thymidylate synthetic pathway (Figure 10A) [71,75]. TS catalyzes the conversion of dUMP to dTMP yielding dihydrofolate, DHF. DHFR uses the coenzyme NADPH to convert DHF, to tetrahydrofolate, THF, for the biosynthesis of purines, thymidylic acid and some amino acids. DeMott et al. 2017 [26] examined the effect of ribosomes on the activity of these enzymes.

The activity of TS increased with increasing ribosome concentration (Figure 10B). In the presence of 0.5 μM ribosomes V_{max} increased ~20-fold, substrate binding affinity decreased ~20-fold and a K_I of 3.9 ± 0.5 μM was resolved (Table 2). The kinetic profile was characteristic of uncompetitive substrate inhibition in which TS-ribosome quinary interactions increased the catalytic rate and promoted substrate inhibition (Figure 10C). In the presence of 0.5 μM ribosomes, DHFR displayed a ~20% decrease in V_{max} and a 10-fold decrease in substrate binding affinity and a kinetic profile consistent with the ribosome acting as a competitive inhibitor (Table 2; Figure 10D). The reduced activity may be due to the DHFR-ribosome interface blocking or altering DHF and/or NADPH binding sites, and/or NADPH binding to ribosomes lowering the concentration of free NADPH available for DHFR catalysis. Indeed, NADPH was shown to bind specifically to ribosomes with a dissociation constant of 4.5 ± 1.5 μM (Figure 10E).

The results suggest a possible mechanism through which ribosome-mediated quinary structural interactions act to reduce cellular levels of dUMP (Figure 10A). Ribosome suppression of DHFR activity lowers the intracellular concentration of THF, which is converted into Me-THF. The decrease in Me-THF concentration reduces the ability of TS to utilize the substrate resulting in a buildup of dUMP (Figure 10A). However, ribosomal-dependent enhancement of TS activity (Figure 10B) increases the catalytic rate allowing the mutagenic substrate to be metabolized. This shows the potential for ribosomes to regulate cellular processes through compensatory adaptations of functional linkages.

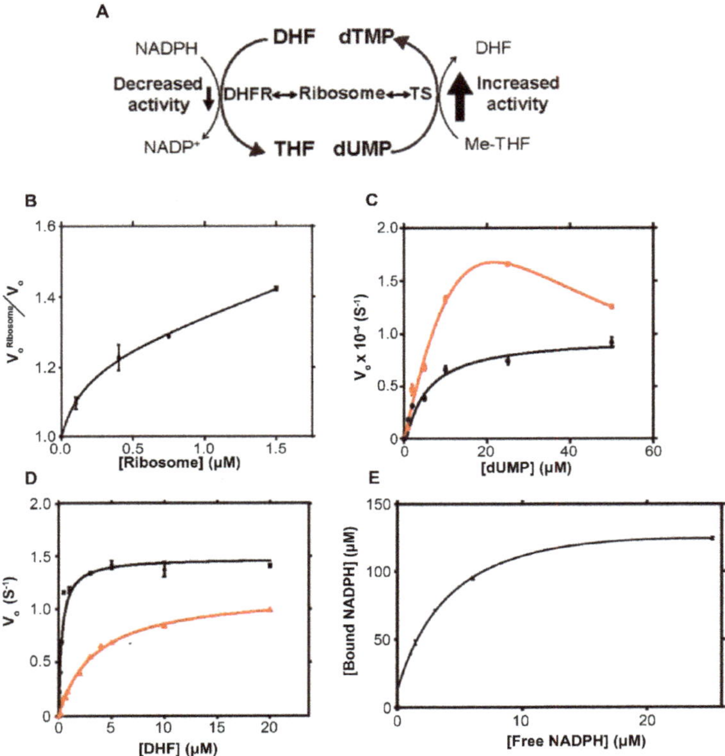

Figure 10. Ribosomes modulate TS and DHFR enzymatic activities. (**A**) Function linkage between TS and DHFR in the thymidylate synthetic pathway (**B**) Increase in TS activity with increasing ribosome concentration. (**C**) Kinetic activity profile for TS without (black) and with (red) 0.5 µM ribosome. (**D**) Kinetic activity profile for DHFR without (black) and with (red) 0.5 µM ribosome. (**E**) NADPH binding to ribosomes. The concentration of ribosomes was 1 µM. Figure is adapted from DeMott et al. (2017) [26].

7. Discussion

The cytosol of an *E. coli* cell is highly congested containing about 300 mg/mL of macromolecules [76]. Such a high concentration creates an enormous excluded volume through macromolecule crowding, which in turn reduces the concentration of bulk water while simultaneously increasing the concentration of macromolecular and ionic species. The reduced water activity affects equilibria governing hydrophobic and hydrophilic interactions and the solvent shells on protein surfaces. The increase in soluble species, in combination with intermolecular distances less than the typical Debye radius for ion charges [77], i.e., ~0.7 nm, reduce the effects of electrostatic screening, promote electrostatic interactions and inevitably increase the propensity for transient low-affinity interactions.

In this review, we summarized work that identified transient low-affinity protein–RNA interactions, historically called quinary [18] by using in-cell ^{1}H-^{15}N CRINEPT-HMQC-TROSY NMR spectroscopy to overcome the effects of extreme broadening of spectral crosspeaks. Quinary structures are large transient complexes that affect protein stability [78,79] and can modulate ligand binding and protein function. Similar interactions have been detected in highly concentrated cell lysates [16,80,81]. The effects of RNA on peak broadening were reconstituted in vitro using preparations of total RNA from both prokaryotic and eukaryotic cells, and purified ribosomes thus confirming the specificity of the interactions [14,26].

The initial observation of extreme crosspeak broadening in in-cell NMR spectra aspired in vitro studies to attribute the phenomenon to the effects of excluded volume, macromolecular crowding and increased intracellular viscosity [15–17,82–84]. These studies provided useful insight into physical mechanisms for limited spectral broadening but none were able to fully reproduce the effects seen in-cell. The binding of a labeled target protein to a large cellular component is the driving force behind spectral broadening due to the reduction in tumbling rate that accompanies the massive increase in the apparent molecular size of the target, which in turn, affects the magnitude of the NMR signal. These interactions underlie the quinary protein structure and can have a profound influence on the activity of the target and its regulation.

Most cytosolic proteins exhibit activity in the absence of other macromolecular species, i.e., in vitro, requiring only substrates and co-factors. Indeed, observations made under these conditions have provided the basis for understanding and modeling much cellular physiology and metabolism. In-cell the activity resulting from metabolic enzymes and other cytosolic species engaging in quinary interactions originates from a population of free and bound species and their derivative functional linkages. In addition, the increase in ribosome concentration with cell growth [85,86] further modulates the distribution between free and bound protein. Thus, the effect of the ribosome on the net activity is to fine-tune and regulate the metabolism of the cell both directly, as is the case for ADK-ribosome interactions, or indirectly, as shown for Trx-mRNA-ribosome interactions in growing cells.

The micromolar concentration of ribosomes in prokaryotes and eukaryotes [24] virtually assures that the ribosomal-binding interactions described, all of which exhibit micromolar dissociation constants, occur inside live cells. Accordingly, we propose that the ribosome plays a role in organizing metabolism [87] by serving as a hub for concentrating enzymes and metabolites. In actively growing *E. coli*, the fractional volume occupied by fully processed 70S ribosomes is ~0.16 [24] and may increase up to four times, ~0.64, outside the space occupied by the nucleoid [88–90]. When compared to the fraction of space occupied by closely packed hard spheres, 0.74, [91] this implies that *E. coli* ribosomes are tightly packed in the cytosol with the volume available for biological reactions restricted to the "free" spaces delimited by ribosome surfaces [92]. In this manner, the surfaces of the ribosome become the operational milieu for much biological activity. Going forward, further in-cell NMR spectroscopy and models of cellular metabolism that depend on activity gleaned in vitro must consider the inescapable effects of ribosomes on these processes.

Author Contributions: Original Draft Preparation, D.S.B.; Review & Editing, D.S.B. and A.S.; Visualization, L.B.

Funding: This work was supported by NIH grant 2R01 GM085006 to A.S.

Conflicts of Interest: The authors declare no conflict of interest.

References

1. Serber, Z.; Dotsch, V. In-cell NMR spectroscopy. *Biochemistry* **2001**, *40*, 14317–14323. [CrossRef] [PubMed]
2. Sakakibara, D.; Sasaki, A.; Ikeya, T.; Hamatsu, J.; Hanashima, T.; Mishima, M.; Yoshimasu, M.; Hayashi, N.; Mikawa, T.; Walchli, M.; et al. Protein structure determination in living cells by in-cell NMR spectroscopy. *Nature* **2009**, *458*, 102–105. [CrossRef]

3. Inomata, K.; Ohno, A.; Tochio, H.; Isogai, S.; Tenno, T.; Nakase, I.; Takeuchi, T.; Futaki, S.; Ito, Y.; Hiroaki, H.; et al. High-resolution multi-dimensional NMR spectroscopy of proteins in human cells. *Nature* **2009**, *458*, 106–109. [CrossRef]

4. Burz, D.S.; Dutta, K.; Cowburn, D.; Shekhtman, A. Mapping structural interactions using in-cell NMR spectroscopy (STINT-NMR). *Nat. Methods* **2006**, *3*, 91–93. [CrossRef]

5. Theillet, F.X.; Binolfi, A.; Bekei, B.; Martorana, A.; Rose, H.M.; Stuiver, M.; Verzini, S.; Lorenz, D.; van Rossum, M.; Goldfarb, D.; et al. Structural disorder of monomeric alpha-synuclein persists in mammalian cells. *Nature* **2016**, *530*, 45–50. [CrossRef]

6. Banci, L.; Barbieri, L.; Bertini, I.; Luchinat, E.; Secci, E.; Zhao, Y.; Aricescu, A.R. Atomic-resolution monitoring of protein maturation in live human cells by NMR. *Nat. Chem. Biol.* **2013**, *9*, 297–299. [CrossRef] [PubMed]

7. Ogino, S.; Kubo, S.; Umemoto, R.; Huang, S.; Nishida, N.; Shimada, I. Observation of NMR Signals from Proteins Introduced into Living Mammalian Cells by Reversible Membrane Permeabilization Using a Pore-Forming Toxin, Streptolysin O. *J. Am. Chem. Soc.* **2009**, *131*, 10834–10835. [CrossRef] [PubMed]

8. Maldonado, A.Y.; Burz, D.S.; Reverdatto, S.; Shekhtman, A. Fate of Pup inside the Mycobacterium proteasome studied by in-cell NMR. *PLoS ONE* **2013**, *8*, e74576. [CrossRef]

9. Hough, L.E.; Dutta, K.; Sparks, S.; Temel, D.B.; Kamal, A.; Tetenbaum-Novatt, J.; Rout, M.P.; Cowburn, D. The molecular mechanism of nuclear transport revealed by atomic-scale measurements. *Elife* **2015**, *4*. [CrossRef] [PubMed]

10. Wall, K.P.; Hough, L.E. In-Cell NMR within Budding Yeast Reveals Cytoplasmic Masking of Hydrophobic Residues of FG Repeats. *Biophys. J.* **2018**, *115*, 1690–1695. [CrossRef]

11. Selenko, P.; Serber, Z.; Gadea, B.; Ruderman, J.; Wagner, G. Quantitative NMR analysis of the protein G B1 domain in Xenopus laevis egg extracts and intact oocytes. *Proc. Natl. Acad. Sci. USA* **2006**, *103*, 11904–11909. [CrossRef]

12. Luchinat, E.; Gianoncelli, A.; Mello, T.; Galli, A.; Banci, L. Combining in-cell NMR and X-ray fluorescence microscopy to reveal the intracellular maturation states of human superoxide dismutase 1. *Chem. Commun.* **2015**, *51*, 584–587. [CrossRef]

13. Maldonado, A.Y.; Burz, D.S.; Shekhtman, A. In-cell NMR spectroscopy. *Prog. Nucl. Magn. Reson. Spectrosc.* **2011**, *59*, 197–212. [CrossRef]

14. Majumder, S.; Xue, J.; DeMott, C.M.; Reverdatto, S.; Burz, D.S.; Shekhtman, A. Probing protein quinary interactions by in-cell nuclear magnetic resonance spectroscopy. *Biochemistry* **2015**, *54*, 2727–2738. [CrossRef]

15. Ye, Y.; Wu, Q.; Zheng, W.; Jiang, B.; Pielak, G.J.; Liu, M.; Li, C. Quantification of size effect on protein rotational mobility in cells by ^{19}F NMR spectroscopy. *Anal. Bioanal. Chem.* **2018**, *410*, 869–874. [CrossRef]

16. Ye, Y.; Liu, X.; Zhang, Z.; Wu, Q.; Jiang, B.; Jiang, L.; Zhang, X.; Liu, M.; Pielak, G.J.; Li, C. ^{19}F NMR spectroscopy as a probe of cytoplasmic viscosity and weak protein interactions in living cells. *Chemistry* **2013**, *19*, 12705–12710. [CrossRef]

17. Crowley, P.B.; Chow, E.; Papkovskaia, T. Protein interactions in the Escherichia coli cytosol: An impediment to in-cell NMR spectroscopy. *Chembiochem* **2011**, *12*, 1043–1048. [CrossRef]

18. McConkey, E.H. Molecular evolution, intracellular organization, and the quinary structure of proteins. *Proc. Natl. Acad. Sci. USA* **1982**, *79*, 3236–3240. [CrossRef]

19. Wirth, A.J.; Gruebele, M. Quinary protein structure and the consequences of crowding in living cells: Leaving the test-tube behind. *Bioessays* **2013**, *35*, 984–993. [CrossRef]

20. Cohen, R.D.; Pielak, G.J. A cell is more than the sum of its (dilute) parts: A brief history of quinary structure. *Protein Sci.* **2017**, *26*, 403–413. [CrossRef]

21. Theillet, F.X.; Binolfi, A.; Frembgen-Kesner, T.; Hingorani, K.; Sarkar, M.; Kyne, C.; Li, C.; Crowley, P.B.; Gierasch, L.; Pielak, G.J.; et al. Physicochemical properties of cells and their effects on intrinsically disordered proteins (IDPs). *Chem. Rev.* **2014**, *114*, 6661–6714. [CrossRef]

22. Luchinat, E.; Banci, L. A Unique Tool for Cellular Structural Biology: In-cell NMR. *J. Biol. Chem.* **2016**, *291*, 3776–3784. [CrossRef]

23. Serber, Z.; Corsini, L.; Durst, F.; Dotsch, V. In-cell NMR spectroscopy. *Methods Enzymol.* **2005**, *394*, 17–41.

24. Milo, R.; Philips, R. *Cell Biology by the Numbers*, 1st ed.; Garland Science: New York, NY, USA, 2015; p. 393.

25. Milo, R.; Jorgensen, P.; Moran, U.; Weber, G.; Springer, M. BioNumbers—The database of key numbers in molecular and cell biology. *Nucleic Acids Res.* **2010**, *38*, D750–D753. [CrossRef]

26. DeMott, C.M.; Majumder, S.; Burz, D.S.; Reverdatto, S.; Shekhtman, A. Ribosome mediated quinary interactions modulate in-cell protein activities. *Biochemistry* **2017**, *56*, 4117–4126. [CrossRef]

27. Castello, A.; Fischer, B.; Frese, C.K.; Horos, R.; Alleaume, A.M.; Foehr, S.; Curk, T.; Krijgsveld, J.; Hentze, M.W. Comprehensive Identification of RNA-Binding Domains in Human Cells. *Mol. Cell* **2016**, *63*, 696–710. [CrossRef]

28. Castello, A.; Fischer, B.; Eichelbaum, K.; Horos, R.; Beckmann, B.M.; Strein, C.; Davey, N.E.; Humphreys, D.T.; Preiss, T.; Steinmetz, L.M.; et al. Insights into RNA biology from an atlas of mammalian mRNA-binding proteins. *Cell* **2012**, *149*, 1393–1406. [CrossRef]

29. Simsek, D.; Tiu, G.C.; Flynn, R.A.; Byeon, G.W.; Leppek, K.; Xu, A.F.; Chang, H.Y.; Barna, M. The Mammalian Ribo-interactome Reveals Ribosome Functional Diversity and Heterogeneity. *Cell* **2017**, *169*, 1051–1065. [CrossRef]

30. Beckmann, B.M.; Horos, R.; Fischer, B.; Castello, A.; Eichelbaum, K.; Alleaume, A.M.; Schwarzl, T.; Curk, T.; Foehr, S.; Huber, W.; et al. The RNA-binding proteomes from yeast to man harbour conserved enigmRBPs. *Nat. Commun.* **2015**, *6*, 10127. [CrossRef]

31. Feig, M.; Yu, I.; Wang, P.H.; Nawrocki, G.; Sugita, Y. Crowding in Cellular Environments at an Atomistic Level from Computer Simulations. *J. Phys. Chem. B* **2017**, *121*, 8009–8025. [CrossRef]

32. Yu, I.; Mori, T.; Ando, T.; Harada, R.; Jung, J.; Sugita, Y.; Feig, M. Biomolecular interactions modulate macromolecular structure and dynamics in atomistic model of a bacterial cytoplasm. *Elife* **2016**, *5*. [CrossRef]

33. Schavemaker, P.E.; Smigiel, W.M.; Poolman, B. Ribosome surface properties may impose limits on the nature of the cytoplasmic proteome. *Elife* **2017**, *6*. [CrossRef]

34. Liu, B.; Poolman, B.; Boersma, A.J. Ionic Strength Sensing in Living Cells. *ACS Chem. Biol.* **2017**, *12*, 2510–2514. [CrossRef]

35. Majumder, S.; DeMott, C.M.; Reverdatto, S.; Burz, D.S.; Shekhtman, A. Total Cellular RNA Modulates Protein Activity. *Biochemistry* **2016**, *55*, 4568–4573. [CrossRef]

36. Breindel, L.; DeMott, C.; Burz, D.S.; Shekhtman, A. Real-Time In-Cell Nuclear Magnetic Resonance: Ribosome-Targeted Antibiotics Modulate Quinary Protein Interactions. *Biochemistry* **2018**. [CrossRef]

37. Ercikan-Abali, E.A.; Banerjee, D.; Waltham, M.C.; Skacel, N.; Scotto, K.W.; Bertino, J.R. Dihydrofolate reductase protein inhibits its own translation by binding to dihydrofolate reductase mRNA sequences within the coding region. *Biochemistry* **1997**, *36*, 12317–12322. [CrossRef]

38. Chu, E.; Takimoto, C.H.; Voeller, D.; Grem, J.L.; Allegra, C.J. Specific binding of human dihydrofolate reductase protein to dihydrofolate reductase messenger RNA in vitro. *Biochemistry* **1993**, *32*, 4756–4760. [CrossRef]

39. Brunn, N.D.; Dibrov, S.M.; Kao, M.B.; Ghassemian, M.; Hermann, T. Analysis of mRNA recognition by human thymidylate synthase. *Biosci. Rep.* **2014**, *34*, e00168. [CrossRef]

40. Bertrand, K.; Reverdatto, S.; Burz, D.S.; Zitomer, R.; Shekhtman, A. Structure of proteins in eukaryotic compartments. *J. Am. Chem. Soc.* **2012**, *134*, 12798–12806. [CrossRef]

41. Li, C.; Wang, G.F.; Wang, Y.; Creager-Allen, R.; Lutz, E.A.; Scronce, H.; Slade, K.M.; Ruf, R.A.; Mehl, R.A.; Pielak, G.J. Protein ^{19}F NMR in Escherichia coli. *J. Am. Chem. Soc.* **2010**, *132*, 321–327. [CrossRef]

42. Danielsson, J.; Inomata, K.; Murayama, S.; Tochio, H.; Lang, L.; Shirakawa, M.; Oliveberg, M. Pruning the ALS-associated protein SOD1 for in-cell NMR. *J. Am. Chem. Soc.* **2013**, *135*, 10266–10269. [CrossRef]

43. Barnes, C.O.; Pielak, G.J. In-cell protein NMR and protein leakage. *Proteins* **2011**, *79*, 347–351. [CrossRef]

44. Wuthrich, K. *NMR of Proteins and Nucleic Acids*; John Wiley&Sons: New York, NY, USA, 1986; p. 293.

45. Cavanagh, J.F.; Fairbrother, W.J.; Palmer, A.G.; Rance, M.; Skelton, N.J. *Protein NMR Spectroscopy*; Academic Press: New York, NY, USA, 2007.

46. Li, C.; Charlton, L.M.; Lakkavaram, A.; Seagle, C.; Wang, G.; Young, G.B.; Macdonald, J.M.; Pielak, G.J. Differential dynamical effects of macromolecular crowding on an intrinsically disordered protein and a globular protein: Implications for in-cell NMR spectroscopy. *J. Am. Chem. Soc.* **2008**, *130*, 6310–6311. [CrossRef]

47. Pielak, G.J.; Li, C.; Miklos, A.C.; Schlesinger, A.P.; Slade, K.M.; Wang, G.F.; Zigoneanu, I.G. Protein nuclear magnetic resonance under physiological conditions. *Biochemistry* **2009**, *48*, 226–234. [CrossRef]

48. Serber, Z.; Keatinge-Clay, A.T.; Ledwidge, R.; Kelly, A.E.; Miller, S.M.; Dotsch, V. High-resolution macromolecular NMR spectroscopy inside living cells. *J. Am. Chem. Soc.* **2001**, *123*, 2446–2447. [CrossRef]

49. Riek, R.; Pervushin, K.; Wuthrich, K. TROSY and CRINEPT: NMR with large molecular and supramolecular structures in solution. *Trends Biochem. Sci.* **2000**, *25*, 462–468. [CrossRef]

50. Pervushin, K.V.; Wider, G.; Wuthrich, K. Single Transition-to-single Transition Polarization Transfer (ST2-PT) in ^{15}N-^{1}H TROSY. *J. Biomol. NMR* **1998**, *12*, 345–348. [CrossRef]

51. Riek, R.; Wider, G.; Pervushin, K.; Wuthrich, K. Polarization transfer by cross-correlated relaxation in solution NMR with very large molecules. *Proc. Natl. Acad. Sci. USA* **1999**, *96*, 4918–4923. [CrossRef]

52. Goldman, M. Interference effects in the relaxation of a pair of unlike 1/2 nuclei. *J. Magn. Reson.* **1984**, *60*, 437–452. [CrossRef]

53. Shekhtman, A.; Ghose, R.; Goger, M.; Cowburn, D. NMR structure determination and investigation using a reduced proton (REDPRO) labeling strategy for proteins. *FEBS Lett.* **2002**, *524*, 177–182. [CrossRef]

54. Riek, R.; Fiaux, J.; Bertelsen, E.B.; Horwich, A.L.; Wuthrich, K. Solution NMR techniques for large molecular and supramolecular structures. *J. Am. Chem. Soc.* **2002**, *124*, 12144–12153. [CrossRef]

55. Schimmel, P.R.; Cantor, C.R. *Biophysical Chemistry: Part II: Techniques for the Study of Biologycal Structure and Function*; W.H. Freeman: New York, NY, USA, 1980; p. 650.

56. Segur, J.; Oberstar, H. Viscosity of Glycerol and its Aqueous Solutions. *Ind. Eng. Chem.* **1951**, *43*, 2117–2120. [CrossRef]

57. Burz, D.S.; Dutta, K.; Cowburn, D.; Shekhtman, A. In-cell NMR for protein-protein interactions (STINT-NMR). *Nat. Protoc.* **2006**, *1*, 146–152. [CrossRef]

58. Burz, D.S.; DeMott, C.M.; Aldousary, A.; Dansereau, S.; Shekhtman, A. Quantitative Determination of Interacting Protein Surfaces in Prokaryotes and Eukaryotes by Using In-Cell NMR Spectroscopy. *Methods Mol. Biol.* **2018**, *1688*, 423–444. [CrossRef]

59. Burz, D.S.; Shekhtman, A. The STINT-NMR method for studying in-cell protein-protein interactions. *Curr. Protoc. Protein Sci.* **2010**, *61*, 17. [CrossRef]

60. Majumder, S.; DeMott, C.M.; Burz, D.S.; Shekhtman, A. Using singular value decomposition to characterize protein-protein interactions by in-cell NMR spectroscopy. *ChemBioChem* **2014**, *15*, 929–933. [CrossRef]

61. Beis, I.; Newsholme, E.A. The contents of adenine nucleotides, phosphagens and some glycolytic intermediates in resting muscles from vertebrates and invertebrates. *Biochem. J.* **1975**, *152*, 23–32. [CrossRef]

62. Henzler-Wildman, K.A.; Thai, V.; Lei, M.; Ott, M.; Wolf-Watz, M.; Fenn, T.; Pozharski, E.; Wilson, M.A.; Petsko, G.A.; Karplus, M.; et al. Intrinsic motions along an enzymatic reaction trajectory. *Nature* **2007**, *450*, 838–844. [CrossRef]

63. Aden, J.; Verma, A.; Schug, A.; Wolf-Watz, M. Modulation of a pre-existing conformational equilibrium tunes adenylate kinase activity. *J. Am. Chem. Soc.* **2012**, *134*, 16562–16570. [CrossRef]

64. Schneider, D.A.; Gourse, R.L. Relationship between growth rate and ATP concentration in *Escherichia coli*: A bioassay for available cellular ATP. *J. Biol. Chem.* **2004**, *279*, 8262–8268. [CrossRef]

65. Freeberg, M.A.; Han, T.; Moresco, J.J.; Kong, A.; Yang, Y.C.; Lu, Z.J.; Yates, J.R.; Kim, J.K. Pervasive and dynamic protein binding sites of the mRNA transcriptome in Saccharomyces cerevisiae. *Genome Biol.* **2013**, *14*, R13. [CrossRef] [PubMed]

66. Castello, A.; Hentze, M.W.; Preiss, T. Metabolic Enzymes Enjoying New Partnerships as RNA-Binding Proteins. *Trends Endocrinol. Metab.* **2015**, *26*, 746–757. [CrossRef] [PubMed]

67. Baltz, A.G.; Munschauer, M.; Schwanhausser, B.; Vasile, A.; Murakawa, Y.; Schueler, M.; Youngs, N.; Penfold-Brown, D.; Drew, K.; Milek, M.; et al. The mRNA-bound proteome and its global occupancy profile on protein-coding transcripts. *Mol. Cell* **2012**, *46*, 674–690. [CrossRef]

68. Vazquez-Laslop, N.; Mankin, A.S. Context-Specific Action of Ribosomal Antibiotics. *Annu. Rev. Microbiol.* **2018**, *72*, 185–207. [CrossRef]

69. Futterman, S. Enzymatic reduction of folic acid and dihydrofolic acid to tetrahydrofolic acid. *J. Biol. Chem.* **1957**, *228*, 1031–1038.

70. Schnell, J.R.; Dyson, H.J.; Wright, P.E. Structure, dynamics, and catalytic function of dihydrofolate reductase. *Annu. Rev. Biophys. Biomol. Struct.* **2004**, *33*, 119–140. [CrossRef]

71. Carreras, C.W.; Santi, D.V. The catalytic mechanism and structure of thymidylate synthase. *Annu. Rev. Biochem.* **1995**, *64*, 721–762. [CrossRef]

72. Sinev, M.A.; Sineva, E.V.; Ittah, V.; Haas, E. Towards a mechanism of AMP-substrate inhibition in adenylate kinase from Escherichia coli. *FEBS Lett.* **1996**, *397*, 273–276. [CrossRef]

73. Cleland, W.W. Substrate inhibition. *Methods Enzymol.* **1979**, *63*, 500–513.

74. Kovermann, M.; Grundstrom, C.; Sauer-Eriksson, A.E.; Sauer, U.H.; Wolf-Watz, M. Structural basis for ligand binding to an enzyme by a conformational selection pathway. *Proc. Natl. Acad. Sci. USA* **2017**, *114*, 6298–6303. [CrossRef]

75. Fox, J.T.; Stover, P.J. Folate-mediated one-carbon metabolism. *Vitam. Horm.* **2008**, *79*, 1–44. [CrossRef]

76. Goodsell, D. *The Machinery of Life*; Springer: New York, NY, USA, 2009; p. 167.

77. Benedek, G.B.; Villars, F.M.H. *Physics with Illustrative Examples from Medicine and Biology: Electricity and Magnetism*; Springer: New York, NY, USA, 2000.

78. Danielsson, J.; Mu, X.; Lang, L.; Wang, H.; Binolfi, A.; Theillet, F.X.; Bekei, B.; Logan, D.T.; Selenko, P.; Wennerstrom, H.; et al. Thermodynamics of protein destabilization in live cells. *Proc. Natl. Acad. Sci. USA* **2015**, *112*, 12402–12407. [CrossRef]

79. Monteith, W.B.; Cohen, R.D.; Smith, A.E.; Guzman-Cisneros, E.; Pielak, G.J. Quinary structure modulates protein stability in cells. *Proc. Natl. Acad. Sci. USA* **2015**, *112*, 1739–1742. [CrossRef]

80. Kyne, C.; Ruhle, B.; Gautier, V.W.; Crowley, P.B. Specific ion effects on macromolecular interactions in Escherichia coli extracts. *Protein Sci.* **2015**, *24*, 310–318. [CrossRef]

81. Barbieri, L.; Luchinat, E.; Banci, L. Protein interaction patterns in different cellular environments are revealed by in-cell NMR. *Sci. Rep.* **2015**, *5*, 14456. [CrossRef]

82. Kyne, C.; Crowley, P.B. Short Arginine Motifs Drive Protein Stickiness in the Escherichia coli Cytoplasm. *Biochemistry* **2017**, *56*, 5026–5032. [CrossRef]

83. Li, C.; Pielak, G.J. Using NMR to distinguish viscosity effects from nonspecific protein binding under crowded conditions. *J. Am. Chem. Soc.* **2009**, *131*, 1368–1369. [CrossRef]

84. Li, C.; Wang, Y.; Pielak, G.J. Translational and rotational diffusion of a small globular protein under crowded conditions. *J. Phys. Chem. B* **2009**, *113*, 13390–13392. [CrossRef]

85. Jensen, K.F.; Pedersen, S. Metabolic growth rate control in Escherichia coli may be a consequence of subsaturation of the macromolecular biosynthetic apparatus with substrates and catalytic components. *Microbiol. Rev.* **1990**, *54*, 89–100.

86. Gourse, R.L.; Gaal, T.; Bartlett, M.S.; Appleman, J.A.; Ross, W. rRNA transcription and growth rate-dependent regulation of ribosome synthesis in Escherichia coli. *Annu. Rev. Microbiol.* **1996**, *50*, 645–677. [CrossRef]

87. Srere, P.A. Complexes of sequential metabolic enzymes. *Annu. Rev. Biochem.* **1987**, *56*, 89–124. [CrossRef]

88. Sanamrad, A.; Persson, F.; Lundius, E.G.; Fange, D.; Gynna, A.H.; Elf, J. Single-particle tracking reveals that free ribosomal subunits are not excluded from the Escherichia coli nucleoid. *Proc. Natl. Acad. Sci. USA* **2014**, *111*, 11413–11418. [CrossRef]

89. Bakshi, S.; Siryaporn, A.; Goulian, M.; Weisshaar, J.C. Superresolution imaging of ribosomes and RNA polymerase in live Escherichia coli cells. *Mol. Microbiol.* **2012**, *85*, 21–38. [CrossRef]

90. Fisher, J.K.; Bourniquel, A.; Witz, G.; Weiner, B.; Prentiss, M.; Kleckner, N. Four-dimensional imaging of *E. coli* nucleoid organization and dynamics in living cells. *Cell* **2013**, *153*, 882–895. [CrossRef]

91. Conway, J.H.; Sloane, N.J.A. *Sphere Packings, Latticies, and Groups*; Springer: Berlin, Germany, 1999; p. 681.

92. Ortiz, J.O.; Forster, F.; Kurner, J.; Linaroudis, A.A.; Baumeister, W. Mapping 70S ribosomes in intact cells by cryoelectron tomography and pattern recognition. *J. Struct. Biol.* **2006**, *156*, 334–341. [CrossRef]

International Journal of
Molecular Sciences

Article

In Situ Monitoring of Bacteria under Antimicrobial Stress Using ^{31}P Solid-State NMR

Sarah A. Overall [1,2], Shiying Zhu [1], Eric Hanssen [3], Frances Separovic [1,*] and Marc-Antoine Sani [1,*]

[1] School of Chemistry, Bio21 Institute, University of Melbourne, Victoria 3010, Australia; soverall@ucsc.edu (S.A.O.); shiyingz2@student.unimelb.edu.au (S.Z.)
[2] Chemistry & Biochemistry Department, UC Santa Cruz, CA 95064, USA
[3] Advanced Microscopy Facility and Department of Biochemistry & Molecular Biology, Bio21 Institute, University of Melbourne, Victoria 3010, Australia; ehanssen@unimelb.edu.au
* Correspondence: fs@unimelb.edu.au (F.S.); msani@unimelb.edu.au (M.-A.S.); Tel.: +61-3-9035-7539 (F.S.)

Received: 30 November 2018; Accepted: 1 January 2019; Published: 6 January 2019

Abstract: In-cell NMR offers great insight into the characterization of the effect of toxins and antimicrobial peptides on intact cells. However, the complexity of intact live cells remains a significant challenge for the analysis of the effect these agents have on different cellular components. Here we show that ^{31}P solid-state NMR can be used to quantitatively characterize the dynamic behaviour of DNA within intact live bacteria. Lipids were also identified and monitored, although ^{31}P dynamic filtering methods indicated a range of dynamic states for phospholipid headgroups. We demonstrate the usefulness of this methodology for monitoring the activity of the antibiotic ampicillin and the antimicrobial peptide (AMP) maculatin 1.1 (Mac1.1) against Gram-negative bacteria. Perturbations in the dynamic behaviour of DNA were observed in treated cells, which indicated additional mechanisms of action for the AMP Mac1.1 not previously reported. This work highlights the value of ^{31}P in-cell solid-state NMR as a tool for assessing the antimicrobial activity of antibiotics and AMPs in bacterial cells.

Keywords: antimicrobial peptide; live cell; solid-state NMR; DNA; lipid membrane

1. Introduction

The emergence of bacterial strains resistant to many important front line antibiotics has fuelled the urgency for which new efficacious antimicrobial agents are being developed. Antimicrobial peptides (AMPs) are produced by many organisms to protect against infections and are an attractive class of antimicrobial agent due to their low cytotoxicity and low occurrence of bacterial resistance [1,2]. AMPs are typically characterized by their cationic charge and amphipathic structure [3] conferring greater activity towards negatively charged membranes, typical of bacteria, while neutral membranes, characteristic of eukaryotes, are less affected [4,5], thus, conferring specificity towards bacterial membranes. The targeted disruption of bacterial membranes by AMPs can occur by several mechanisms from pore formation to detergent like solubilization of membranes [6–8]. However, it is unclear what the contribution of membrane disruption is to the bactericidal properties of AMPs. The complex architecture of the bacterial membrane could modulate the mode of action of AMPs in situ, altering the threshold of inhibitory concentrations. Thus, there are still unknowns in understanding how AMPs interact with bacteria and, in particular, how bacteria respond to AMP activity in situ.

Maculatin 1.1 (Mac1.1) is an AMP derived from the skin secretions of the Australian tree frog, *Litoria genimaculata*. Mac1.1 is a typical cationic AMP showing low µM activity against Gram-positive bacteria and moderate activity against Gram-negative bacteria [9,10]. It is proposed to act via a pore-forming mechanism [11]; however, the discrepancy in threshold concentration between bacterial species was recently hypothesized as a consequence of bacterial responses to AMP induced stress instead of peptide activity alone [12]. Activity of AMPs against non-membrane targets have also been documented for other AMPs, displaying direct interaction with bacterial DNA, inhibition of protein synthesis and destabilization of the cell wall [13]. These effects often synergize with membrane permeabilization properties and so understanding how AMPs interact with various cellular components is critical for the rational design of synergistic AMP therapies. Reproducing the complex double membrane structure of Gram-negative bacteria in a model system is challenging given the differential lipid composition of the inner and outer membranes separated by peptidoglycan. To address this, we present an initial in-cell portrayal of intact and viable *E. coli* bacteria using ^{31}P solid-state NMR spectroscopy. In-cell NMR is a fast-developing method to investigate the structure and function of biomolecules in their native environment [14,15]. We have applied this approach to study the effect of the well-characterized antibiotic ampicillin and the AMP Mac1.1 on *E. coli* bacteria. We demonstrate the feasibility of using ^{31}P NMR to identify nucleic acids (NAs), i.e., DNA and RNA, and lipid components in intact bacteria. Detailed characterization of NA phosphodiester bond dynamics was obtained with only a qualitative assessment of lipid structure and dynamics due to signal overlap and a broad dynamic range. Using this methodology, we show that ampicillin and Mac1.1 have significant effects on the dynamics of NAs resulting in increased motional averaging of the NA signal. These effects could be related to reactive oxygen species (ROS) mediated stress as the presence of thiourea, a ROS scavenger, could partly ameliorate the effects of ampicillin on the NA spectrum.

2. Results

2.1. Solid-State ^{31}P NMR Experiments on Intact E. coli Bacteria

Phosphorous containing molecules provide useful markers for understanding how AMPs interact with bacteria as many biomolecules contain phosphorus, particularly, phospholipids and DNA. Utilizing ^{31}P solid-state NMR, we investigated the effect of antimicrobial agents on live *E. coli*. Given the use of small NMR rotors holding less than 100 μL of sample, hydration, oxygen availability and nutrient supply could severely limit the sample integrity and, therefore, data quality over the course of an experiment. To determine the length of time in which live cell samples are sufficiently stable for NMR analysis, we assessed bacterial survival at different time points during an NMR experiment at 30 and 10 °C under static condition, i.e., without magic angle spinning (MAS). Samples were prepared by growing bacteria to an OD_{600} of 0.5, media components were removed by washing in isosmotic salt buffer prior to packing the cell slurry into a 4 mm zirconia MAS rotor and immediately acquiring NMR spectra. Cell survival was quantified by the number of colony forming units (CFUs) in an aliquot of cells taken from the rotor at various time points. Bacterial viability was remarkably stable over a 4 h period given the high cell density (>10^{10} CFU/mL). After 12 h the number of CFU decreased by 70% at 30 °C while cell survival was greatly improved by maintaining the sample at 10 °C, in which CFU declined by only 33% over the same period (Figure 1A). ^{31}P static spectra collected at 10 and 30 °C showed a superposition of lineshapes, indicative of a wide range of ^{31}P environments (Figure 1B,C). After 12 h at 30 °C there was a significant reduction in spectral quality, concomitantly with reduced cell survival (Figure 1A,B). Most notably the intensity of the peak at −11 ppm disappeared and most features merged into a broad isotropic signal, indicative of fast molecular reorientation on the ^{31}P NMR time scale, suggesting a significant loss of cellular integrity and hence sample degradation. However at 10 °C, no drastic change was observed, even after 12 h, and so subsequent experiments were performed for up to 6 h at 10 °C.

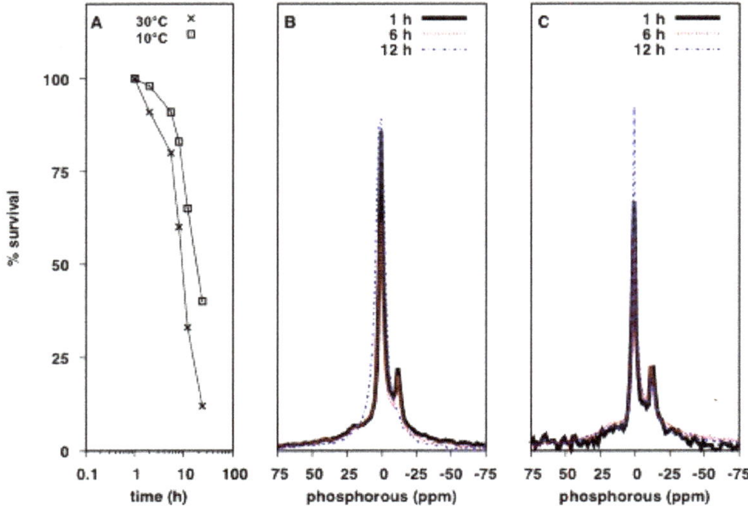

Figure 1. Effect of temperature on bacterial survival over the course of a ^{31}P static NMR experiment: (**A**) Bacterial survival over the course of an NMR experiment. 500 mL of *E. coli* culture, grown to an OD$_{600}$ of 0.5 was packed into a 4 mm zirconia MAS rotor as a cell slurry. Colony forming units were determined by taking 5 μL of cell slurry and performing serial dilutions at each time point, followed by plating onto LB agar and enumerating the number of colonies. Data is from a single experiment with CFU dilutions performed in triplicate. Error bars indicate the standard error of the mean per triplicate. (**B**) ^{31}P direct excitation spectra of the same samples measured in A at 30 °C; and (**C**) at 10 °C. Spectra were taken 1 h (black solid line), 6 h (red dotted line) and 12 h (blue dashed line) after packing.

2.2. Qualitative Analysis of ^{31}P Lineshapes in Intact Bacteria

In order to gain meaningful insights into the effect of antimicrobial agents on the molecular architecture of intact bacteria, we assessed whether different phosphorus containing biomolecules could be distinguished from one another in live bacteria by ^{31}P solid-state NMR. The static ^{31}P lineshape is defined by the width of the powder pattern due to the chemical shift anisotropy (CSA, defined using the Haeberlen convention: $\delta_{zz}-\delta_{iso}$) and asymmetry parameter eta (η), which are modulated by the molecular environments and molecular dynamics of the ^{31}P tensor. The use of ^{1}H to ^{31}P cross-polarization (^{31}P CP) selectively filters out ^{31}P signals of fast rotating molecules and depends on the contact time. Using a relatively short contact time of 1.5 ms, a very broad ^{31}P signal with a span of ca. 230 ppm was observed in bacteria (Figure 2A). This is typical of solid or rigid molecules with little motional averaging of the ^{31}P tensor. The immobilized nature of the signal identified was surprising considering the cells analyzed were well above freezing temperature and were not only intact but also fully hydrated. We postulated that NAs would be the most likely biomolecules to maintain a rigid-like structure, particularly in the form of condensed genomic DNA or large mRNA that can reach megaDalton size [16]. The distinction between DNA and RNA cannot be achieved under static NMR conditions and, since the DNA/RNA mass ratio in *E. coli* varies from ca. 1:3 at early growth phase to ca. 1:15 at late growth phase [17], the overall NA signal was used.

Comparison of ^{31}P CP spectra of purified DNA with that of bacteria revealed identical lineshapes and a comparable chemical shift span or powder pattern of ca. 200 ppm, indicating DNA could be the primary source of this signal. Furthermore, ^{31}P CP analysis of MLVs composed of *E. coli* total lipid extract, providing an expected ^{31}P static lineshape for phospholipids, showed a chemical shift span of ca. 54 ppm (Figure 2A), typical of a gel phase lipid bilayer. Interestingly, the peak intensity at −11 ppm, indicative of fast-axially reorienting lipids, was also visible by CP, indicating that lipids can be detected in this dynamic range, although the dominance of the NA signal as this contact time precludes analysis. Together the data strongly indicates that broad ^{31}P signals in live cells arise due to NAs.

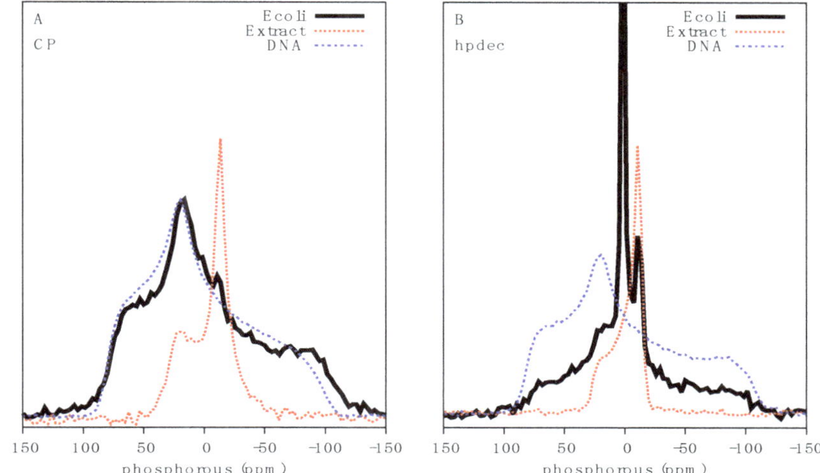

Figure 2. Characterization of bacterial DNA by ^{31}P solid-state NMR: (**A**) ^{31}P CP spectra using a contact time of 1.5 ms; and (**B**) Direct ^{31}P excitation spectra of *E. coli* (black solid line), MLVs made of *E. coli* total lipid extract (red dotted line) and dry genomic DNA (blue dash-dotted line). All spectra were acquired at 10 °C.

Analysis of direct excitation ^{31}P spectra of bacteria, which detects all ^{31}P signals irrespective of molecular motion, revealed a convolution of several powder pattern lineshapes (Figure 2B). An intense isotropic signal centered at 1.8 ppm indicates the presence of fast re-orienting ^{31}P containing molecules and likely arises from a wide range of molecules from phosphorylated soluble proteins, ATP, ADP and inorganic phosphate species. This component accounts for 38% of the spectral area. The underlying rigid NA signal is also visible, albeit significantly reduced relative to the intensity of isotropic ^{31}P dominating the spectrum. A clear powder pattern indicative of a lipid bilayer was also visible as recapitulated by the lipid extracts, typical of a fluid phase phospholipid bilayer (Figure 2B). The outer and inner lipid membranes, which are mainly composed of lipopolysaccharide, phosphatidylethanolamine and phosphatidylglycerol, were difficult to distinguish due to the complexity of the sample. An attempt to deconvolute the ^{31}P spectrum of *E. coli* was made using a three-component system accounting for NA, lipid and an isotropic component. Initial CSA and η values already described for dry DNA and lipid extracts were used as component estimates. Lineshape analysis, using the TopSpin solid lineshape fitting tool, of *E. coli* CP spectra gave a CSA of −109 ppm with an η value of 0.52 and an isotropic chemical shift at 1.28 ppm. Lineshape analysis of the fluid phase lipid component of direct excitation spectra yielded a CSA of 25.2 ppm (overall span of ~45 ppm), η value of 0.12 and isotropic chemical shift of −0.38 ppm, while the isotropic component of the spectra showed a chemical shift of 1.6 ppm. Overall, the fit is in good agreement with the values determined for purified DNA and lipid extract.

2.3. Characterisation of the Effects of Antimicrobial Agents on Bacteria

In order to gain a sense for how AMPs might affect the NMR spectrum of intact bacteria, we first assessed the effects of the well characterized antibiotic ampicillin, whose mechanism of action and cellular effects are well known, including inhibition of cell wall synthesis and damage by reactive oxygen species (ROS) [18,19]. We treated 500 mL *E. coli* cultures with 20 µg/mL ampicillin (minimum inhibition concentration (MIC) of 4 µg/mL against *E. coli* [20]) for 1 h prior to preparation for NMR spectroscopy. Analysis of intensity normalized CP spectra shows a clear reduction in the chemical shift span of the NA signal in ampicillin treated *E. coli*, decreasing from ca. 225 to ca. 150 ppm (Figure 3A). The increased noise in the spectrum is also indicative of a significant reduction in absolute signal intensity observed in ampicillin treated cultures due to cell growth inhibition and cell death. The region of greatest intensity also shifted from 20 to 1 ppm indicating a reduction in tensor asymmetry and increased motional averaging. In the direct excitation spectra, the intensity at -11 ppm due to axially reorienting lipids was significantly reduced (Figure 3B) consistent with membrane perturbations. As ampicillin is not known to directly interact with bacterial DNA, we hypothesized that the reduction in NA chemical shift span could be related to stress induced ROS production [19], possibly damaging NAs through the breaking of phosphodiester bonds resulting in increased ^{31}P motion and thus a reduction in linewidth (Figure 3A). To test this we cultured *E. coli* in the presence of ampicillin and thiourea, a powerful ROS scavenger. Provision of thiourea slightly ameliorated the reduction in chemical shift span of ^{31}P CP spectra and direct excitation spectra (Figure 3). Signals indicative of fast-axially reorienting lipids (signal at -11 ppm) were only slightly recovered in the presence of thiourea, indicating that thiourea treatment could not completely perturb or negate the activity of ampicillin.

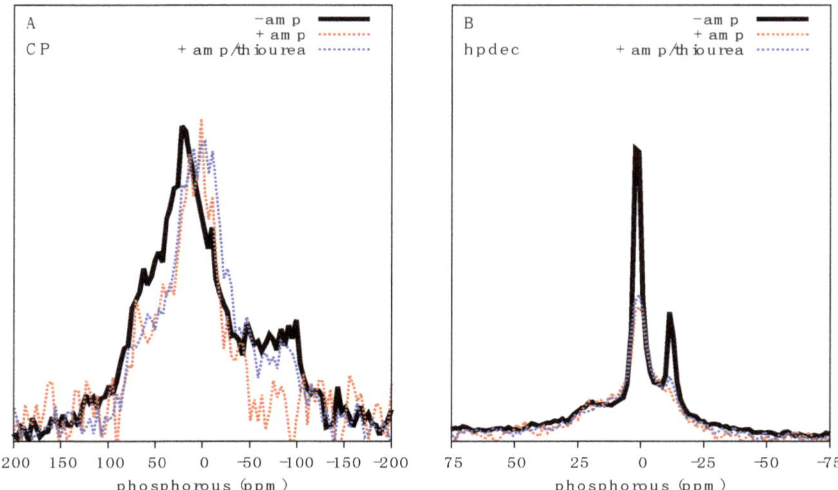

Figure 3. Characterization of the effect of ampicillin on *E. coli*: Bacteria grown to OD_{600} = 0.5 were treated with 20 µg/mL ampicillin for 1 h prior to NMR analysis. (**A**) ^{31}P CP spectra at 1.5 ms contact time, and (**B**) ^{31}P direct excitation spectra of *E. coli* (black solid line), ampicillin treated *E. coli* (red dotted line) and ampicillin and thiourea treated *E. coli* (blue dotteded line). Spectra have been scaled to facilitate lineshape comparisons.

We next assessed the impact of the AMP, Mac1.1 on *E. coli* by ^{31}P solid-state NMR. *E. coli* cultures were treated with sub-MIC concentrations of Mac1.1 (MIC of order 100 µg/mL or greater against *E. coli*, [12]) for 30 min prior to NMR analysis. At 10 µM Mac1.1, a loss of signal intensity at the edges of the ^{31}P CP spectrum was observed (Figure 4A,B). Increasing the amount of Mac1.1 to 25 µM intensified the lineshape alterations. The results indicate a change in NA dynamics with an increase in the proportion of molecules undergoing fast reorientation as observed following ampicillin treatment. Direct excitation ^{31}P spectra showed a similar increase in fast-reorienting ^{31}P molecules significantly reducing the signal intensity of axially reorienting lipids, indicative of membrane disruption (Figure 4C,D).

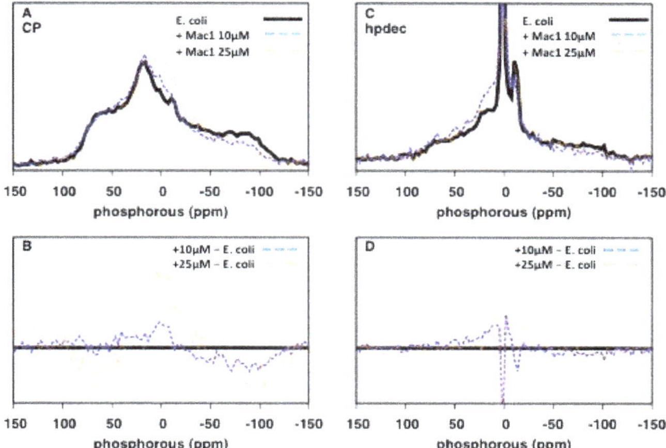

Figure 4. Mac1.1 perturbs NA packing: (**A**) ^{31}P 1.5 ms contact time CP spectra; and (**C**) ^{31}P direct excitation spectra of *E. coli* (black solid line), in the presence of 10 μM Mac1.1 (blue dashed line) and 25 μM Mac1.1 (orange dotted line). (**B**) Difference between CP spectra, and (**D**) difference between direct excitation spectra of *E. coli* and in the presence of 10 μM Mac1.1 (blue dashed line) or in the presence of 25 μM Mac1.1 (orange dotted line).

Quantitative analysis of CP spectra with various contact times revealed a reduction in $T_{1\rho}$ values following Mac1.1 treatment, dropping from 4.8 to 3.3 ms (Figure 5D), indicative of increased molecular motion in the μs regime and therefore a reduction in molecular order and stability, consistent with the qualitative changes observed in the lineshape and chemical shift span. Given the similarities between Mac1.1 and ampicillin treatment on ^{31}P spectra of NAs, we similarly treated *E. coli* with Mac1.1 in the presence of thiourea to determine the contribution of ROS damage to Mac1.1 mediated effects on the spectra. However, rather than being protective, thiourea treatment caused catastrophic cell death when combined with Mac1.1 precluding any analysis due to almost complete loss of signal from the death of the cell culture.

The full effect of Mac1.1 on membrane integrity was difficult to determine by NMR given the spectral overlap both in the direct excitation and CP spectra. To further confirm the disrupting effect of Mac1.1 on *E. coli* membranes we utilized electron microscopy. *E. coli* cultured in the presence of 25 μM Mac1.1 showed a collapse of the periplasmic space with no distinction between the inner and outer membranes compared to untreated bacteria (Figure 6). This effect could be replicated in the presence of 0.1% Triton-X, indicating that the physiological disruption of the periplasmic space is similar to detergent treatment. Furthermore, Mac1.1 altered DNA morphology within the cytosolic space, causing the appearance of increased condensation compared to untreated *E. coli* while Triton-X had the opposite effect, dispersing the DNA (Figure 6). Comparison of ^{31}P CP spectra of Mac1.1 and Triton-X treated cells revealed significant qualitative changes that were more pronounced in the case of Mac1.1. Signal intensity was severely reduced in both spectra due to cell death, impeding lineshape comparison. However, the spectra indicate that Triton-X has a minimal effect on the ^{31}P CP lineshape, unlike the AMP (Figures 5C and 6). Collectively, the data suggests that disruption of the ^{31}P NA signal by Mac1.1 was not due to unraveling of genomic DNA and thus the AMP effect on NAs is distinct from its effect on membranes rather than as a consequence of membrane disruption.

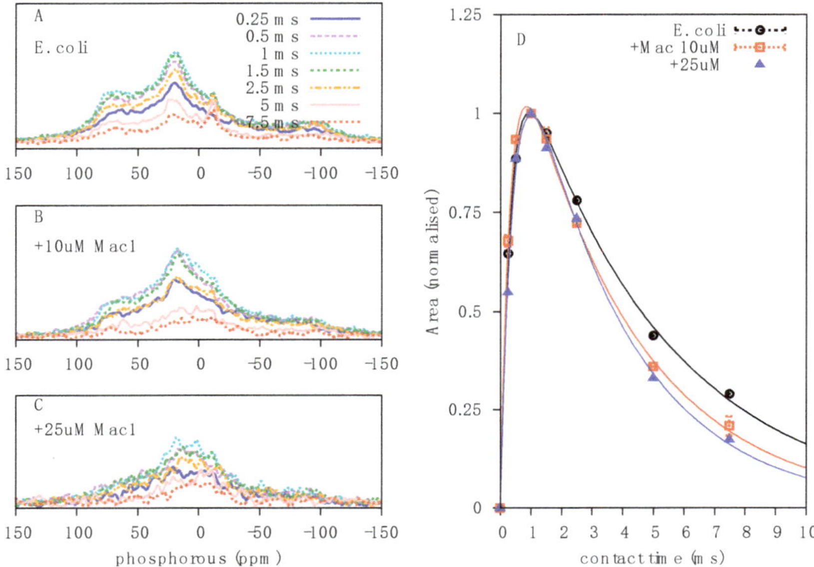

Figure 5. Mac1.1 increased μs motions in NAs: Contact time array in ^{31}P CP spectra of: (**A**) peptide-free *E. coli*; (**B**) in the presence of 10 μM Mac1.1; and (**C**) in the presence of 25 μM Mac1.1. The contact times used were (solid blue) 0.25 ms, (dashed purple) 0.5 ms, (dotted cyan) 1 ms, (dash-dotted green) 1.5 ms, (dash-dotted-dotted gold) 2.5 ms, (short-dotted pink) 5 ms and (short-dashed red) 7.5 ms. (**D**) Integral versus contact time plot with fit according to Equation (1) of (circle red line) peptide-free *E. coli*, (square cyan) in the presence of 10 μM Mac1.1 and (triangle blue) in the presence of 25 μM Mac1.1.

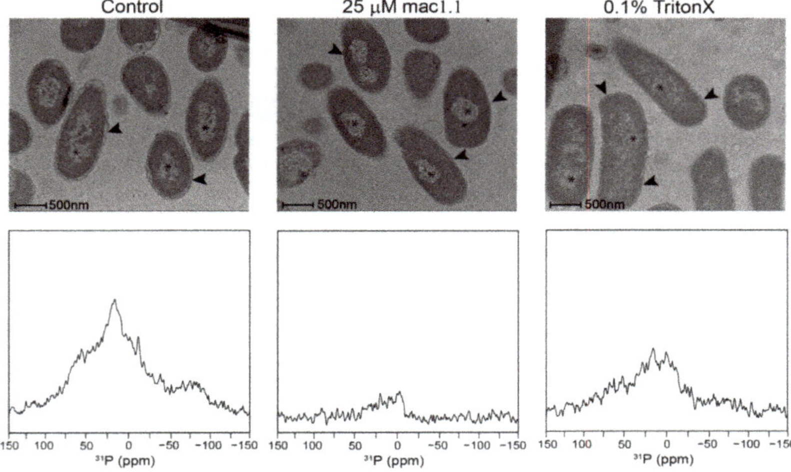

Figure 6. DNA morphology modulation and intermembrane collapse induced by Mac1.1: Upper panels show electron micrographs of *E. coli* (**left panel**), Mac1.1 treated (**middle panel**) and 0.1% triton-X treated (**right panel**). Asterisks indicates genomic DNA, arrow heads indicate the periplasmic space. *E. coli* were prepared as previously described for NMR experiments but washed in phosphate buffered saline (PBS) prior to preparation for transmission electron microscopy. Lower panel shows the comparable ^{31}P CP NMR spectra performed at 30 °C.

3. Discussion

In-cell NMR provides a means for understanding the molecular effects of antimicrobial agents in a cellular context. While recent developments have focused on structure and folding of cytosolic proteins [21], little has been done to investigate the potential for using ^{31}P as a molecular probe of cellular integrity. We have shown that phospholipids and NA can be readily distinguished in intact cells by ^{31}P solid-state NMR and used effectively to probe the effects of antimicrobial agents on cellular integrity. Both ampicillin and Mac1.1 enhanced motional averaging of the ^{31}P tensor of NAs, which may be partly attributed to possible ROS mediated damage of the NA backbone. Membrane integrity was also affected by both agents but due to spectral overlap could only be assessed through the intensity of the peak attributable to axially reorienting lipids.

The ^{31}P spectral width observed in bacteria was greater than expected for live cells, particularly the presence of apparently rigid or immobilized molecules displaying a lineshape covering the full ^{31}P CSA. Comparison with purified DNA corroborated the molecular identity of this signal and is also consistent with ^{31}P lineshapes observed in viral particles in which the vast majority of ^{31}P atoms can be attributed to NA in the form of viral genomic DNA or RNA [22]. Treatment of *E. coli* with either Mac1.1 or ampicillin resulted in similar lineshape alterations, with a reduction in signal intensity at the edges of the spectrum and a shift in spectral density towards 0 ppm, which collectively indicate increased dynamics of the ^{31}P tensor of NA and confirmed by relaxation analysis of this signal in Mac1.1 treated bacteria. The partial recovery of this signal in ampicillin treated bacteria with thiourea needs further investigations and could involve ROS as ampicillin is widely known to induce ROS and is an important contributor to its effectiveness as an antimicrobial [19]. Since Mac1.1 is unable to create large pores [11], such as formed by cytolysins, it is unlikely that NAs are released from the intracellular space, causing the spectral changes observed. This was further confirmed by electron microscopy (EM) where Mac1.1 did not cause severe membrane disruption but rather increased DNA compaction. While the mechanism driving this morphological change in DNA is unclear, it is interesting to note that the CP spectra change in a similar way upon Mac1.1 and Triton-X treatment, showing different changes in DNA compaction by EM. If phosphodiester bond breakage occurred, then increased bond rotation of the phosphate would increase the isotropic nature of the signal and lead to reduced CP spectral intensity. In the absence of bond breakage, increased compaction would be expected to reduce $T_{1\rho}$, resulting in greater signal intensities at short contact times, which was not observed in the semi-quantitative CP analysis (Figure 5D). Disruption of the DNA or RNA macromolecular structure, however, could also reduce the ^{31}P powder pattern linewidth as would an increase in motion due to unraveled genomic DNA. Averaging of the ^{31}P tensor would still be restricted by the phosphodiester bond dynamics and but may not change ^{31}P dynamics to the same extent as expected in the case of bond breakage. Direct excitation spectra make it clear that this is a dynamic change, at least in the case of Mac1.1, in which spectral area is equivalent between treated and untreated cells. Breakage and fragmentation of NAs remains a possibility that requires further investigation using a combination of gel electrophoresis and mass spectrometry.

In the presence of thiourea, Mac1.1 treatment caused catastrophic cell death, resulting in almost complete destruction of the bacteria and dramatic loss of ^{31}P NMR signal, preventing analysis. This was a powerful demonstration of the potential for combining sub-MIC concentrations of pore forming AMPs with other antimicrobial agents having different activity such that a normally protective anti-oxidant could become a powerful antimicrobial in the presence of AMPs. It is becoming increasingly clear that synergy between antimicrobial agents has the potential to raise the efficacy of existing antibiotics, even against strains that demonstrate resistance to frontline antibiotics like vancomycin [23,24]. This finding also highlights the potential for Mac1.1 to be used as a drug delivery system, particularly for other powerful antimicrobial agents that may exhibit poor membrane solubility and transport.

The majority of AMPs have a strong affinity for lipid membranes, in particular negatively charged membranes, which confers specificity towards anionic prokaryotic membranes. Mac1.1 is a typical membrane-active peptide that has been shown to possess greater affinity for anionic membranes compared to neutral membranes and to form pores in bacterial membranes. A recent study showed that Mac1.1 alters *E. coli* membrane integrity well below the MIC and thus suggested that bacteria can cope differently with membrane damage [12], which would explain the difference in MIC between bacterial species. Both ampicillin and Mac1.1 destabilized membrane integrity, increasing the contribution of an isotropic component to the overall spectrum and a loss of signal from axially reorienting lipids. Disruption of the membrane by Mac1.1 was further visualized by electron microscopy with the appearance of a collapsed periplasmic space.

Observation of phospholipid resonances was difficult due to signal overlap in both direct excitation and cross-polarization spectra, precluding detailed analysis of the effects of either antimicrobial agent on lipid dynamics and membrane structure. Furthermore, the observation of axially reorienting lipids in CP spectra in addition to direct excitation spectra inferred at a wide dynamic range of phospholipid dynamics in intact live bacteria. Lipids that form close interactions with membrane proteins would be expected to exhibit reduced dynamics and, therefore, able to be visualized in CP spectra at short contact times. The CSA of these powder patterns would reveal the dynamic range of these lipids, though this was not possible due to the dominance of the rigid NA signal. This range of dynamics in which phospholipid headgroups appear to exist in live cells potentially makes observation of lipids in cells difficult by ^{31}P NMR as dynamic filters cannot be effectively used, at least in the absence of lipid specific labelling schemes [25]. However, disentanglement of heterogeneous dynamics in mixed lipid systems using MAS techniques [26] could be used in future to study live bacteria, providing bacteria integrity is preserved during MAS experiments.

4. Materials and Methods

4.1. Materials

Maculatin 1.1 (GLFGVLAKVAAHVVPAIAEHF-NH2; MW 2145) at >95% purity was made at the Bio21 Institute peptide synthesis facility (Melbourne, Australia). The peptide was washed in 5 mM HCl and lyophilized over-night to remove residual trifluoroacetic acid [27]. *Escherichia coli* total lipid extract was purchased from Avanti Polar Lipids (Alabaster, AL, USA) and used without further purification. Peptone, yeast extract, piperazine-N,N'-bis(2-ethanesulfonic acid) (PIPES) and sodium chloride were purchased from Sigma (Castle Hill, Australia).

4.2. Bacterial Growth

BL21 (DE3) *E. coli* was obtained from New England Biolabs (Ipswich, MA, USA.). Innoculation of an agar plate was made and incubated overnight at 37 °C. A single colony was transferred into a Luria broth (LB) culture flask and grown to an OD_{600} of 0.5 at 37 °C and under 250 rpm orbital shaking. Bacterial treatments were begun at OD_{600} of 0.5. Thiourea was used at a final concentration of 150 mM, ampicillin was used at 20 μg/mL and Mac1.1 at 10 and 25 μM where indicated. After the times indicated, the bacterial suspension was pelleted at 4000 rpm, washed twice with PIPES buffer (100 mM PIPES, 50 mM NaCl, pH 7.2). The supernatant was removed and the cell pellet was centrifuged down into a 4 mm Bruker NMR rotor (Billerica, MA, USA) at 1000 rpm.

4.3. Bacterial Survival

Survival of *E. coli* was determined by taking 5 μL of *E. coli* suspension from a packed rotor and performing serial dilutions in LB media, plated out onto agar and grown overnight at 37 °C. Colony forming units (CFU) were determined by counting the number of colonies. The percentage of remaining CFUs at various time points was calculated by: (number of CFU at 0 h)/(number of CFU at the indicated time points) [20].

4.4. E. coli Total Lipid Extract Sample Preparation

Multilamellar vesicles (MLVs) were formed using *E. coli* total lipid extract, which was resuspended in a large excess of Milli-Q water with or without Mac1.1, thoroughly homogenized and freeze-dried overnight. The dry powder was then resuspended in PIPES buffer at 65% (*w/w*) hydration and freeze-thawed three times prior to packing into a 4 mm MAS NMR rotor.

4.5. Solid-State NMR Spectroscopy

^{31}P NMR experiments were performed on a DNP-NMR 400 MHz Bruker Avance spectrometer at a frequency of 161.5 MHz. A 4 mm triple resonance probe was used in a double resonance mode. 62.5 kHz direct excitation pulse and 55 kHz Hartmann-Hahn ^{1}H to ^{31}P cross-polarisation (^{31}P CP) experiments with 10% ramp on proton spin-lock were used under 31.25 kHz ^{1}H SPINAL decoupling scheme. The recycle delay was obtained from ^{1}H and ^{31}P saturation recovery experiments, and set at 6 s (~5 T1). Typically, 512 scans and 3k scans were acquired for direct and 1.5 ms CP excitation experiments, respectively, except for the CP contact-time array and processed with 8k zero-filling and line-broadening from 50 to 500 Hz were used.

The CP spectra were integrated using the Topspin 3.5 software (Bruker, Billerica, MA, USA) and the spectrum area versus contact time experiments fitted according to [28]:

$$A(t) = A_0 \left(1 - T_{H\text{-}P}/T^H_{1\rho}\right) - 1 \left[\exp(-t/T^H_{1\rho}) - \exp(-t/T_{H\text{-}P})\right] \tag{1}$$

where $A(t)$ is the spectral area, A_0 is proportional to an equilibrium Bloch magnetization, $T_{H\text{-}P}$ is the CP time constant between ^{1}H and ^{31}P, and $T^H_{1\rho}$ is the ^{1}H spin-lattice relaxation in the rotating frame.

4.6. Electron Microscopy

Bacteria were grown in LB media until OD_{600} = 0.5 then treated with 25 µM Mac1.1 dissolved in water, water alone or 0.1% Triton-X for 30 min. The cells were then collected by centrifugation and washed twice with phosphate buffered saline prior to preparation for imaging. The samples were fixed with 1.5% formaldehyde, 1% glutaraldehyde in 0.1 M sodium cacodylate rinsed and post fixed in 1% osmium tetroxide in 0.15 M sodium cacodylate the serially dehydrate in ethanol and embedded in epoxy resin (Procure 812). Seventy nanometer sections were cut and stained with uranyl acetate and lead citrate. Samples were observed at 200 kV using a FEI Tecnai F30 (Hillsboro, OR, USA) equipped with a Gatan US1000 digital camera (Pleasanton, CA, USA).

Author Contributions: Conceptualization, S.A.O., F.S. and M.-A.S.; Data curation, F.S. and M.-A.S.; Formal analysis, S.A.O., F.S. and M.-A.S.; Funding acquisition, F.S.; Investigation, S.A.O., S.Z., F.S. and M.-A.S.; Methodology, S.A.O., F.S. and M.-A.S.; Project administration, F.S.; Resources, E.H. and F.S.; Supervision, F.S. and M.-A.S.; Writing—original draft, S.A.O. and M.-A.S.; Writing—review & editing, S.A.O., E.H., F.S. and M.-A.S.

Funding: This research was funded by the Australian Research Council, grant number DP 160100959.

Acknowledgments: The authors are grateful for the access to the Bio21 Institute NMR, EM and peptide synthesis facilities.

Conflicts of Interest: The authors declare no conflict of interest.

Abbreviations

NMR	Nuclear Magnetic Resonance
CP	Cross-Polarization
CSA	Chemical Shift Anisotropy
AMP	Antimicrobial Peptide
Mac1.1	Maculatin 1.1
DNA	Deoxyribonucleic Acid
CFU	Colony Forming Unit

References

1. Zasloff, M. Antimicrobial peptides of multicellular organisms. *Nature* **2002**, *415*, 389–395. [CrossRef] [PubMed]
2. Jenssen, H.; Hamill, P.; Hancock, R.E.W. Peptide antimicrobial agents. *Clin. Microbiol. Rev.* **2006**, *19*, 491–511. [CrossRef] [PubMed]
3. Brogden, K.A. Antimicrobial peptides: Pore formers or metabolic inhibitors in bacteria? *Nat. Rev. Microbiol.* **2005**, *3*, 238–250. [CrossRef] [PubMed]
4. Sani, M.A.; Gagne, E.; Gehman, J.D.; Whitwell, T.C.; Separovic, F. Dye-release assay for investigation of antimicrobial peptide activity in a competitive lipid environment. *Eur. Biophys. J. EBJ* **2014**, *43*, 445–450. [CrossRef] [PubMed]
5. Yeaman, M.R.; Yount, N.Y. Mechanisms of antimicrobial peptide action and resistance. *Pharmacol. Rev.* **2003**, *55*, 27–55. [CrossRef] [PubMed]
6. Sani, M.A.; Separovic, F. Progression of NMR studies of membrane-active peptides from lipid bilayers to live cells. *J. Magn. Reson.* **2015**, *253*, 138–142. [CrossRef] [PubMed]
7. Sani, M.A.; Separovic, F. How Membrane-Active Peptides Get into Lipid Membranes. *Acc. Chem. Res.* **2016**, *4*, 1130–1138. [CrossRef]
8. Melo, M.N.; Ferre, R.; Castanho, M.A.R.B. Antimicrobial peptides: Linking partition, activity and high membrane-bound concentrations. *Nat. Rev. Microbiol.* **2009**, *7*, 245–250. [CrossRef]
9. Chia, B.C.S.; Carver, J.A.; Mulhern, T.D.; Bowie, J.H. Maculatin 1.1, an anti-microbial peptide from the Australian tree frog, *Litoria genimaculata*—Solution structure and biological activity. *Eur. J. Biochem.* **2000**, *267*, 1894–1908. [CrossRef]
10. Rozek, T.; Waugh, R.J.; Steinborner, S.T.; Bowie, J.H.; Tyler, M.J.; Wallace, J.C. The maculatin peptides from the skin glands of the tree frog *Litoria genimaculata*: A comparison of the structures and antibacterial activities of maculatin 1.1 and caerin 1.1. *J. Pept. Sci.* **1998**, *4*, 111–115. [CrossRef]
11. Sani, M.A.; Whitwell, T.C.; Gehman, J.D.; Robins-Browne, R.M.; Pantarat, N.; Attard, T.J.; Reynolds, E.C.; O'Brien-Simpson, N.M.; Separovic, F. Maculatin 1.1 disrupts *Staphylococcus aureus* lipid membranes via a pore mechanism. *Antimicrob. Agents Chemother.* **2013**, *57*, 3593–3600. [CrossRef] [PubMed]
12. Sani, M.A.; Henriques, S.T.; Weber, D.; Separovic, F. Bacteria May Cope Differently from Similar Membrane Damage Caused by the Australian Tree Frog Antimicrobial Peptide Maculatin 1.1. *J. Biol. Chem.* **2015**, *290*, 19853–19862. [CrossRef] [PubMed]
13. Scocchi, M.; Mardirossian, M.; Runti, G.; Benincasa, M. Non-Membrane Permeabilizing Modes of Action of Antimicrobial Peptides on Bacteria. *Curr. Top. Med. Chem.* **2016**, *16*, 76–88. [CrossRef]
14. Freedberg, D.I.; Selenko, P. Live Cell NMR. *Annu. Rev. Biophys.* **2014**, *43*, 171–192. [CrossRef] [PubMed]
15. Burz, D.S.; Dutta, K.; Cowburn, D.; Shekhtman, A. In-cell NMR for protein-protein interactions (STINT-NMR). *Nat. Protoc.* **2006**, *1*, 146–152. [CrossRef] [PubMed]
16. Majumder, S.; DeMott, C.M.; Reverdatto, S.; Burz, D.S.; Shekhtman, A. Total Cellular RNA Modulates Protein Activity. *Biochemistry* **2016**, *55*, 4568–4573. [CrossRef] [PubMed]
17. Shahab, N.; Flett, F.; Oliver, S.G.; Butler, P.R. Growth rate control of protein and nucleic acid content in Streptomyces coelicolor A3(2) and *Escherichia coli* B/r. *Microbiology* **1996**, *142 Pt 8*, 1927–1935. [CrossRef]
18. Smirnova, G.; Muzyka, N.; Lepekhina, E.; Oktyabrsky, O. Roles of the glutathione- and thioredoxin-dependent systems in the *Escherichia coli* responses to ciprofloxacin and ampicillin. *Arch. Microbiol.* **2016**, *198*, 913–921. [CrossRef]
19. Kohanski, M.A.; Dwyer, D.J.; Hayete, B.; Lawrence, C.A.; Collins, J.J. A common mechanism of cellular death induced by bactericidal antibiotics. *Cell* **2007**, *130*, 797–810. [CrossRef]
20. Andrews, J.M. Determination of minimum inhibitory concentrations. *J. Antimicrob. Chemother.* **2001**, *48*, S5–S16. [CrossRef]
21. Luchinat, E.; Banci, L. In-cell NMR: A topical review. *IUCrJ* **2017**, *4*, 108–118. [CrossRef] [PubMed]
22. Diverdi, J.A.; Opella, S.J. Phosphorus-31 Nuclear Magnetic-Resonance of fd Virus. *Biochemistry* **1981**, *20*, 280–284. [CrossRef] [PubMed]
23. Amani, J.; Barjini, K.A.; Moghaddam, M.M.; Asadi, A. In vitro synergistic effect of the CM11 antimicrobial peptide in combination with common antibiotics against clinical isolates of six species of multidrug-resistant pathogenic bacteria. *Protein Pept. Lett.* **2015**, *22*, 940–951. [CrossRef] [PubMed]

24. Wu, X.Z.; Li, Z.; Li, X.L.; Tian, Y.M.; Fan, Y.Z.; Yu, C.H.; Zhou, B.L.; Liu, Y.; Xiang, R.; Yang, L. Synergistic effects of antimicrobial peptide DP7 combined with antibiotics against multidrug-resistant bacteria. *Drug Des. Dev. Ther.* **2017**, *11*, 939–946. [CrossRef] [PubMed]

25. Laadhari, M.; Arnold, A.A.; Gravel, A.E.; Separovic, F.; Marcotte, I. Interaction of the antimicrobial peptides caerin 1.1 and aurein 1.2 with intact bacteria by ^{2}H solid-state NMR. *BBA-Biomembranes* **2016**, *1858*, 2959–2964. [CrossRef] [PubMed]

26. Sani, M.A.; Separovic, F.; Gehman, J.D. Disentanglement of Heterogeneous Dynamics in Mixed Lipid Systems. *Biophys. J.* **2011**, *100*, L40–L42. [CrossRef] [PubMed]

27. Sani, M.A.; Dufourc, E.J.; Grobner, G. How does the Bax-α1 targeting sequence interact with mitochondrial membranes? The role of cardiolipin. *Biochim. Biophys. Acta* **2009**, *1788*, 623–631. [CrossRef]

28. Kolodziejski, W.; Klinowski, J. Kinetics of cross-polarization in solid-state NMR: A guide for chemists. *Chem. Rev.* **2002**, *102*, 613–628. [CrossRef]

International Journal of
Molecular Sciences

Article

Unambiguous Ex Situ and in Cell 2D ^{13}C Solid-State NMR Characterization of Starch and Its Constituents

Alexandre Poulhazan [1], Alexandre A. Arnold [1], Dror E. Warschawski [1,2] and Isabelle Marcotte [1,*]

[1] Department of Chemistry, Université du Québec à Montréal, Downtown Station, P.O. Box 8888, Montreal, QC H3C 3P8, Canada; poulhazan.alexandre@courrier.uqam.ca (A.P.); arnold.alexandre@uqam.ca (A.A.A.); Dror.Warschawski@ibpc.fr (D.E.W.)

[2] Laboratoire de Biologie Physico-Chimique des Protéines Membranaires, UMR 7099, CNRS, Université Paris Diderot and IBPC, 13 rue Pierre et Marie-Curie, 75005 Paris, France

* Correspondence: marcotte.isabelle@uqam.ca; Tel.: +1-514-987-3000-5015

Received: 7 November 2018; Accepted: 28 November 2018; Published: 30 November 2018

Abstract: Starch is the most abundant energy storage molecule in plants and is an essential part of the human diet. This glucose polymer is composed of amorphous and crystalline domains in different forms (A and B types) with specific physicochemical properties that determine its bioavailability for an organism, as well as its value in the food industry. Using two-dimensional (2D) high resolution solid-state nuclear magnetic resonance (SS-NMR) on ^{13}C-labelled starches that were obtained from *Chlamydomonas reinhardtii* microalgae, we established a complete and unambiguous assignment for starch and its constituents (amylopectin and amylose) in the two crystalline forms and in the amorphous state. We also assigned so far unreported non-reducing end groups and assessed starch chain length, crystallinity and amylose content. Starch was then characterized in situ, i.e., by ^{13}C solid-state NMR of intact microalgal cells. Our in-cell methodology also enabled the identification of the effect of nitrogen starvation on starch metabolism. This work shows how solid-state NMR can enable the identification of starch structure, chemical modifications and biosynthesis in situ in intact microorganisms, eliminating time consuming and potentially altering purification steps.

Keywords: whole cell NMR; magic-angle spinning; 2D INADEQUATE; crystalline and amorphous starch

---

1. Introduction

Starch is, with cellulose, the most abundant carbohydrate that is found in nature. Composed of a polymer of glucose in semicrystalline granules, it is the major form of energy storage for plants [1]. It is also the main energy source in most animal diets and is involved in various food industry processes. Because they are linked with diseases, starches that are resistant to digestive enzymes have been the focus of a growing research emphasis [2]. In addition to its importance in nutrition, starch can be used as an environmentally-friendly low-cost material with no apparent toxicity and can be functionalized for a wide range of applications, such as adhesives, biofilms, biodegradable plastics, pharmacology, etc. Recently, for example, starch was designed for drug delivery using hydroxymethylated material [3].

In the Plantae, starch is stored in grains, structured on different scales, as illustrated in Figure 1. At a micrometer (μm) scale, starch granules are made of amorphous and crystalline regions. Starch is water insoluble, making it easy to purify, and consists mainly of highly branched amylopectin (70 to 85% by weight of short α-1,4 chains with numerous α-1,6-D-glucan linkages) and linear amylose (15 to 30% by weight of long α-1,4-D-glucan with few α-1,6-D-linkage) [4]. Starch is thus a semi-crystalline network of amylose and amylopectin chains stranded into a double-helical structure held by hydrogen bonds. These helices have 6 glucose residues per turn and a pitch of 2.1 nanometer (nm) and can adopt two different crystalline packings identified as the A and B-types [4]. However, many vegetables possess starch grains that contain a mixture of A and B-types categorized as the C-type [4]. These three forms were first identified by X-ray diffraction (XRD) analysis [5]. According to previous works [6,7], the main difference between A and B-type starches is the relative positions of the starch double-stranded helices. In the A-type structure, left-handed parallel-stranded double helices are closely packed into a B2 space group [8,9], while in the B-type, helices are packed into a hexagonal unit cell corresponding to a $P6_1$ space group [10], forming a more hydrated structure.

One of the limitations of the starch industry is the difficulty to control the final product quality. Today, native starch and its derivatives are most frequently characterized by powder X-ray diffraction [4,11], often in combination with SS-NMR (solid-state nuclear magnetic resonance) [12–16]. NMR has the advantage of being a non-destructive technique, thus allowing measurements on intact hydrated samples while providing information on molecular dynamics at an atomic scale, even in the case of complex biomolecules such as starch [4]. The one-dimensional (1D) study of A and B-type starches has already revealed some differences in ^{13}C chemical shifts. In particular, the multiplicity of the C_1 carbon is used to distinguish the two starch crystalline forms [12–16]. The C_2 to C_5 resonances, however, are poorly resolved on 1D spectra, and 2D SS-NMR has not yet been used to unambiguously characterize A-type starch, amorphous- and amylopectin-rich starches. To the best of our knowledge, only pure synthetic amylose has been described by 2D ^{13}C NMR and proved to be in the B-type form [14].

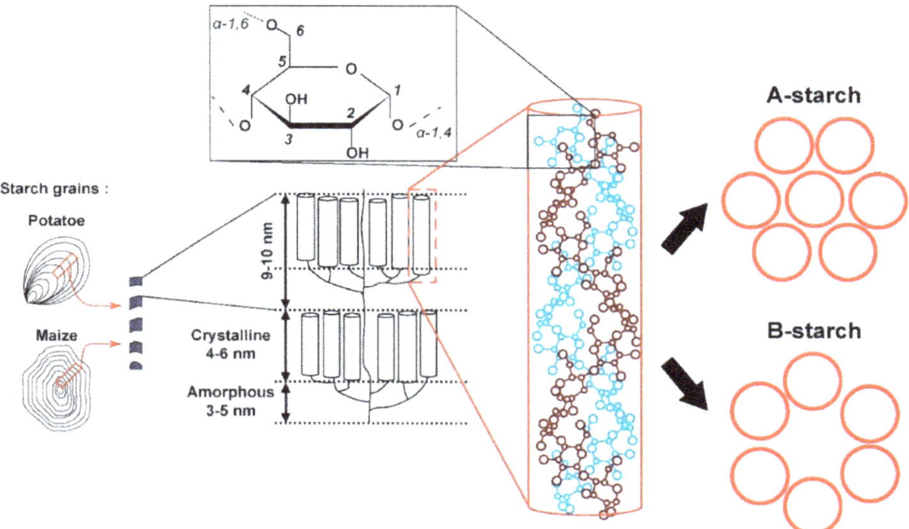

Figure 1. Multiscale representation of starch, from left to right: morphology of starch granules [17], crystalline and amorphous regions, double helices and packing of double helices in crystalline forms A and B.

In this work, we used the microalga *Chlamydomonas reinhardtii*—A photosynthetic model organism with well-known metabolism [18–20], numerous mutants and a fully sequenced genome [21,22]—for starch production and ^{13}C labelling. This microorganism produces starch of the A-type after purification [11,23]. Using amylose-rich, amylopectin-rich and inhibited starch producing strains, we established a full unambiguous high-resolution assignment for all carbons in A, B and amorphous starches—An essential step in the SS-NMR study of starches and their modifications. We assigned highly crystalline amylopectin and poorly crystalline B-type amylose as well as new signals from non-reducing end groups. Beyond its chemical shift and multiplicity, the width and shape of each resonance provided additional data that were interpreted in terms of local vs. global order, correlated disorder, chain length, degree of crystallinity or amylose/amylopectin ratio. Finally, based on these characterizations, we were able to detect the resonances of starch in situ in whole *C. reinhardtii* cells, and to identify the type of starch in the storage grains, as well as the level of crystallinity. This work demonstrates how 2D ^{13}C SS-NMR methodology can prove invaluable for the functional in vivo study of starch in its native environment: the cytosol.

2. Results and Discussion

2.1. 2D ^{13}C SS-NMR Ex Situ Characterization of A and B Types and Amorphous Starch

To characterize starch in situ, we first needed to establish a complete unambiguous ^{13}C assignment of pure A, B and amorphous starches. The A and B forms can respectively be obtained with sufficient high crystallinity using amylopectin-rich and native retrograded starches.

The A form was prepared using starch from the *st 2-1* amylopectin-rich *C. reinhardtii* strain. This strain produces starch with the highest crystallinity—about 71% according to our SS-NMR results (see Table S1). The short chains of amylopectin are reported to form double-helices which readily crystallize in the starch granules [4]. Furthermore, XRD measurements of pure amylopectin (Figure S1A) are typical of highly crystallized A-type starch [11,24]. The B form was obtained by starch retrogradation, as described in the Material and Methods section, from wild-type *C. reinhardtii*, yielding a final crystallinity of *circa* 55% as determined by SS-NMR (Table S1). Finally, amorphous starch was prepared by freeze-drying amylose-rich starch (produced by the *C. reinhardtii sta 3-3* mutant) as described by Paris et al. [15]. The amorphous nature of this sample is confirmed by NMR (about 0% crystallinity, Table S1). Interestingly, the A-type starch remained crystalline after drying, while the crystal structure of B-type starch was destroyed after drying, according to XRD experiments (Figure S1A,B). Water is thus an essential element in the crystal structure of B-type starch, as reported elsewhere [25,26].

As shown in Figure 2, while C_1, C_4 and C_6 carbons are readily resolved on the 1D spectra of types A, B and amorphous starch, carbons 2 to 5 cannot be distinguished. Gidley and Bociek [27] demonstrated that glucose carbons 1 and 4 were more sensitive to starch conformational changes than carbons 2, 3 and 5, showing higher chemical shift dispersion under the variation of the torsion angles of the glycosidic linkage in α-(1,4) glucans. In amorphous starch, the broad distribution of conformations thus leads to a large chemical shift dispersion of the C_1' peak (Figure 2C). The C_1 splitting of the crystalline forms (3 peaks for A and 2 peaks for B) have been explained as resulting from the different space groups adopted by A and B forms [27]. These spatial arrangements lead to three possible environments for carbon 1 in the A form and two for the B form.

A major improvement in resolution can be obtained using 2D methods on ^{13}C labelled material, including whole cells. As we showed in a previous piece of work, microalgae can easily be fully ^{13}C labelled using $NaH^{13}CO_3$ [28]. A higher spectral dispersion provided by the second dimension will reduce the risks of potential overlap between starch and other carbohydrate moieties in situ. The 2D INADEQUATE is an excellent experiment providing unambiguous through-bond connectivities and enhanced resolution. The robustness of this experiment and exquisite sensitivity to conformational differences has been shown in various works on disordered organic materials [29,30]. This experiment has particularly been useful in the study of cellulose [31,32]. Moreover, the double quantum (DQ) dimension provides excellent chemical shift dispersion and the experiment has, thus, also been applied to intact systems [33]. Here, we used the INADEQUATE pulse sequence in combination with proton-to-carbon polarization transfer schemes, such as Cross Polarisation (CP) or the Nuclear Overhauser effect (NOE), which is commonly exploited in solid-state NMR for signal enhancement [28,34].

The 2D INADEQUATE spectra of amylopectin (A-form), retrograded *C. reinhardtii* native starch (B-form) and dry amylose (amorphous) are shown in Figure 3. The net improvement in resolution is sufficient to distinguish all carbons including carbons 2, 3 and 5, which were not unambiguously elucidated before. A complete and unambiguous assignment was thus obtained for the three starch forms, and all spin systems are reported in Table 1. The linewidths vary between ca. 100 Hz and 200 Hz for hydrated starches, thus confirming their well-ordered and dynamic nature. The linewidths of the dry amorphous starch (Figure 3C), on the other hand, can reach 500 Hz, possibly due to the dispersion of conformations and freezing out of motions in this state.

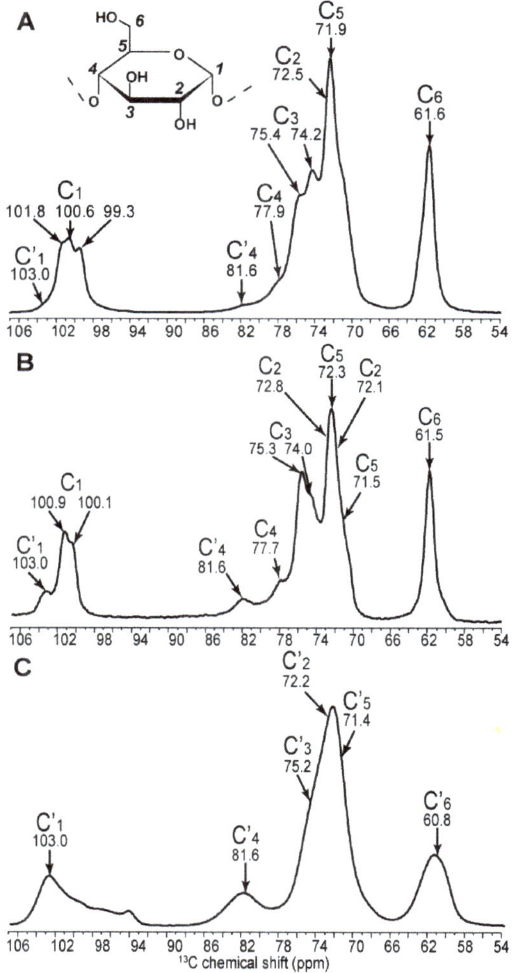

Figure 2. 1D cross-polarisation ^{13}C solid-state NMR (nuclear magnetic resonance) spectra of amylopectin (**A**), retrograded (**B**) and amorphous gelatinized (**C**) starch from *C. reinhardtii*. Assignments are extracted from 2D spectra.

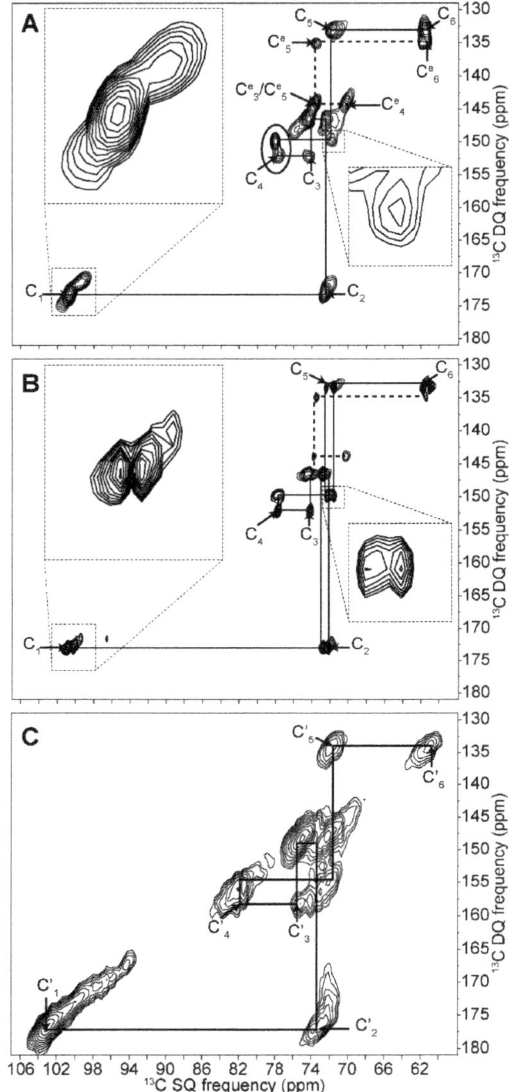

Figure 3. NOE (Nuclear Overhauser Effect)-INADEQUATE 2D SQ/DQ (single quantum/double quantum) ^{13}C solid-state NMR spectra of pure amylopectin A-type starch (**A**), retrograded native *C. reinhardtii* B-type starch (**B**) and amorphous dry native *C. reinhardtii* starch (**C**). Continuous lines are correlation pathways for the main starch constituents (α-1,4 and α-1,6 linked glucans), while dashed lines correspond to non-reducing end groups. Zooms correspond to carbon 1 in C_1–C_2 correlation and to carbon 5 in the C_5–C_4 correlation on the left and right side zooms, respectively. On (**A**), C^e_X designate carbons in non-reducing end-groups. To compare end groups signals between 2D spectra, the intensities are normalized using the C_4 area circled in (**A**).

Table 1. ^{13}C assignments (in ppm) of amylopectin-rich, native retrograded, amylose-rich and native *C. reinhardii* starches [1].

	C1		C2		C3		C4		C5		C6
Amylopectin (A)	101.8	→	73.6								
	100.6	→	72.5	→	74.2	→	77.9	→	71.9	→	61.6
	99.1	→	71.9								
Non reducing end group	100.5	→	72.3	→	73.7	→	70.4	→	73.7	→	61.1
Retrograded (B)	100.9	→	72.1	↘				↗	72.4	→	61.0
					74.1	→	77.7				
	100.1	→	72.8	↗				↘	71.5	→	61.6
Non reducing end group	100.1	→	72.2	→	73.5	→	70.4	→	73.5	→	61.5
Amylose (B)		↗	73.5	↘							
	100.8				74.1	→	77.8	→	71.8	→	61.4
		↘	72.5	↗							
	99.9	→	71.6								
Non reducing end group	100.6	→	72.6	→	73.6	→	70.3	→	73.6	→	61.3
Amorphous	103.0	→	72.7	→	75.4	→	81.5	→	71.5	→	60.4
C. reinhardtii wt	101.5	→	72.1								
Native starch (A)	100.5	→	72.6								
	99.3	→	71.9								
					75.0	→	76.9				
Non reducing end group	100.4	→	72.4	→	73.6	→	70.2	→	73.6	→	61.4
Amorphous	103.0	→	73.0	→	75.3	→	81.7	→	71.5	→	60.2
By-product 1	92.7	→	72.4	→	74.9	→	76.9	→	70.4	→	61.4
By-product 2	96.7	→	74.9								

[1] Amylopectin-rich and amylose-rich starches are purified from st 2-1 and sta 3-1 C. reinhardtii strains, respectively. Correlation pathways are determined using NOE-INADEQUATE except for amorphous starch using CP-INADEQUATE. Arrows represent two correlated carbons in the same spin system.

As seen on the 1D spectrum and as previously reported, the greatest chemical shift differences between starch forms are observed for the C_1 and C_4 carbons [27]. The C_1 value ranges from 99.3 to 101.8 ppm in A-type starch, from 100.1 to 100.9 ppm in B-type and is equal to 103.0 ppm in amorphous starch. The difference in splitting of the C_1 peak between A- and B-types is confirmed and extends to the C_2 carbons for B-type starch, although this difference is lost in the following carbons, except for the C_5 carbon of the B-form. Although chemical shift differences are small, a clear splitting is seen for C_5 in the C_5–C_4 and C_6–C_5 correlation in the B-form spectrum, which is not present in the A-form (Figure 3). The change in C_4 chemical shift between the two crystalline forms is small (0.2 ppm); however, a large 3.9 ppm difference is observed when starch becomes amorphous. Similarly, the difference between C_3 chemical shifts in both crystalline forms is minor (74.2 and 74.1 ppm for A and B forms, respectively), however these can clearly be distinguished from the amorphous chemical shift (75.3 ppm). Overall, differences in chemical shifts between A and B forms are small, indicating that the torsion angles and magnetic environment are very similar. The structures are thus *locally equivalent* and the notable differences that we detected for the C_1 resonances, and which we showed to partially extend to other carbons, result from differences on a longer scale, such as the symmetry of the crystal lattice.

Most importantly, all 2D spectra reveal a clearly distinct spin system (around 100.5, 72.5, 73.5, 70.2, 73.5, 61.3 ppm) that has never been reported in natural starches to the best of our knowledge, most likely because they are almost undetectable on 1D spectra (except for a shoulder near the C_5 carbon). These resonances cannot arise from soluble molecules, such as short carbohydrate oligomers, because they would have been eliminated in the various washing steps of our samples. On the other hand, the chemical shifts of this system of correlated carbons are in excellent agreement with those reported for the end-groups of synthetic alpha dextrins [35] and can, therefore, most certainly be assigned to non-reducing terminal glucose groups.

As such, the intensity of the C^e_4 resonance (Figure 4) cannot allow us to directly measure the percentage of end groups, exact chain length or branching abundance in situ, however the detection of these resonances offers the possibility of obtaining relative values. Absolute quantitative measurements are possible, however they would require prior calibration of resonance intensities with standard molecules. This information can nevertheless help localize chemical reactions in starch and it can also be very useful for the food industry as short amylopectins found in highly resistant starch are valued for their healthful properties [36]. In the context of the development of new rice mutants, for example, following the intensity of the C^e_4 resonance could help predict some of its functional properties and commercial values [37].

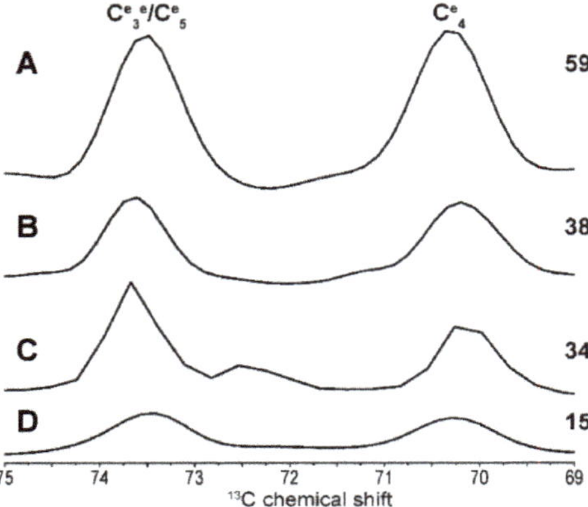

Figure 4. Traces corresponding to the C^e_3/C^e_5–C^e_4 correlation in non-reducing end groups are extracted from 2D spectra with the C_4 area (see Figure 3) arbitrarily set to 100. Amylopectin-rich (**A**), native (**B**), native retrograded (**C**) and amylose-rich (**D**) starches traces are presented. C^e_4 intensities are indicated on the right.

As mentioned previously, we have used signal enhancement schemes, and NOE was usually favoured because it was more efficient than CP, as expected in mobile regions of biomacromolecules. One exception is hydrated amorphous starch, where CP was more efficient and CP-INADEQUATE was preferred (see Figure 5B). In hydrated amorphous starch, Paris and co-workers also used CP and explained its efficiency by the particular nature of proton-to-carbon couplings, where water polarization is transferred by spin diffusion to protons covalently bound to starch carbons in a two ^{1}H reservoir model [16].

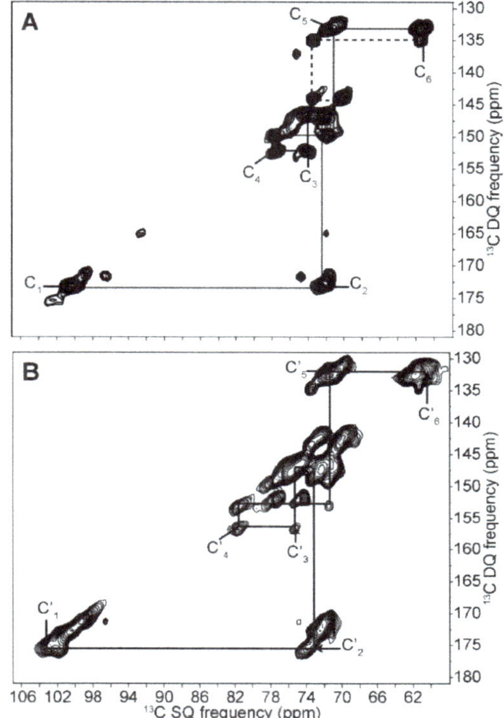

Figure 5. NOE (**A**) and CP (cross-polarization) (**B**) INADEQUATE ^{13}C solid-state NMR spectra of native *C. reinhardtii* A-type starch. Continuous lines link correlated spins in crystalline (**A**) and amorphous (**B**) regions, while dashed lines correspond to non-reducing end groups.

Compared to hydrated starches, the dry amorphous one showed much broader lines with an intrinsic linewidth that was at least twice that of the hydrated forms, as determined by 2D SS-NMR (see Figure S2), and as previously reported in various biological samples [38,39]. The elongated shapes of the lines in the INADEQUATE spectra, which are parallel to a 1:2 axis, reflect a distribution of conformations rather than a mere change in relaxation resulting from the reduced dynamics in the dry form. Moreover, since the resonance pattern of a given carbon is parallel to the previous and the next ones, the distribution of conformations (or disorder) is correlated from one carbon to the adjacent. In other words, the structure is purely amorphous with no regions that are more ordered than others [40]. As soon as starch is hydrated, water acts a plasticizer and chain motions average out the chemical shifts to the values given in Table 1 for hydrated amorphous starch.

To summarize this section, we characterized A and B types and amorphous starch, thus providing the most complete ^{13}C NMR assignment to date, including new chemical shifts ascribed to non-reducing end groups. The widths of 1D lines, as well as the shapes of 2D resonances, are useful indications of molecular order and dynamics. Finally, the ratio of intensities between C_4 and C^e_4 enables the first in situ qualitative assessment of the length of branched amylopectin or linear amylose.

2.2. Differentiating Between Starch Components: Amylose and Amylopectin

Macroscopically, starch has various grain shapes, however with similar architecture consisting of growth rings, blocklets and crystalline-amorphous lamellae. This glycosidic polymer is made of linear and branched sequences, respectively corresponding to amylose and amylopectin. Different ratios of amylopectin/amylose are observed in starch depending on its origin. Pure amylopectin starch has been described as highly crystalline A-type starch [4,41,42], whereas pure amylose starch has been described as having B-type and low crystallinity patterns [43]. Thus, amylose is generally considered to be in the amorphous region of starch granules [4] while amylopectin is in the crystalline region. Nevertheless, many examples show that the amylose/amylopectin ratio is not the only decisive factor determining starch type, crystallinity [44] and digestibility [2].

As expected, starch purification from amylopectin-rich strain *st 2-1* of *C. reinhardtii* leads to highly crystalline A-type starch (for XRD and NMR, see Figures S3 and S4, respectively). Indeed, the 1D ^{13}C SS-NMR experiment shows that the peaks assigned to amorphous C_1 and C_4, respectively at 103 and 81.6 ppm, are nearly absent in amylopectin samples, thus indicating that amylose is at least partially involved in amorphous regions of starch. Furthermore, α-1,6-branched glucose was not detected here because this type of bond has an occurrence of only 5% in amylopectins [4]. As for pure amylose, which is described to poorly crystallize into the B-form [11], chemical shifts are indeed in good agreement with this starch form (Table 1). We here confirm this observation with XRD (Figure S1C) and ^{13}C SS-NMR (Figures S3 and S4) on extracted starch from this strain. Moreover, the same difference could be observed in situ (data not shown). In addition, lines are sharper in the amylose spectrum than in the native and amylopectin rich A-type starch, which is in good agreement with a more amorphous/mobile amylose structure. Interestingly, amylose can also be recrystallised into A- or B-type starch [45], consistent with the higher flexibility of amylose.

This opens the way to exploring the effects of changing amylose/amylopectin ratios related to starch crystallinity and, to some extent, to its type. According to the literature, more amylose is usually more favourable to B-type starch [43], and pure amylose leads to B-type crystals in vitro [46], even if exceptions exist. Amylose/amylopectin ratio and chain length distribution are known to be critical for starch physicochemical properties, which determine their suitability for particular uses. For example, starch films properties [47,48], digestibility and starch water uptake [49] are affected by the amylose/amylopectin ratio. Thus, SS-NMR could be a rapid and efficient tool to help understand these differences.

2.3. In Cell Characterization of C. reinhardtii Starch

As will be seen in this section, our thorough characterization of extracted starch in its various forms enabled us to detect starch in cell and to identify its type and degree of crystallinity. In our previously published work, intense signals were assigned to starch in whole cells of *C. reinhardtii* [28]. Here, we further refined this assignment by exploiting the improved resolution provided by the 2D INADEQUATE experiment. As shown in Figure 6, this experiment can discriminate starch signals from those of other saccharides with the exception of carbons 6 and 3 which overlap with those of the galactolipids and/or present in the cell wall structure. A comparison of the chemical shifts obtained from the 2D ^{13}C INADEQUATE and reported in Table 1 confirms that in *C. reinhardtii* cells, starch crystallizes in the A-form, which is very similar to pure amylopectin, supporting the high proportion of amylopectin in this starch (see XRD and NMR results in Figures S1A,D and S5).

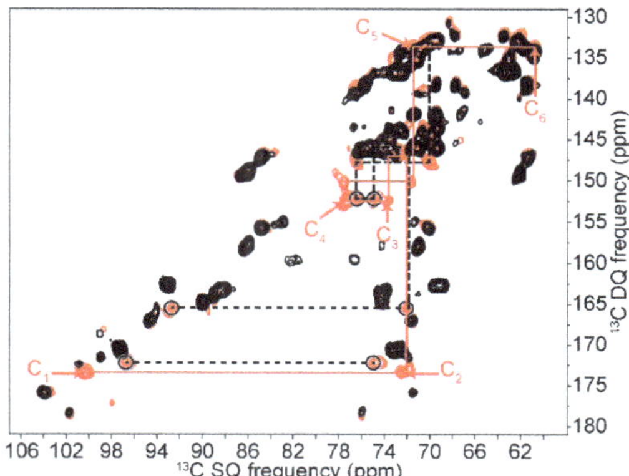

Figure 6. NOE-INADEQUATE of whole cell *C. reinhardtii wt* (red) and *sta 6-1* starchless (black) strains. Continuous lines indicate correlation pathways corresponding to A-type crystalline starch. Dashed lines link circled resonances from the additional correlations discussed in the text.

Our approach can be applied to other microorganisms, however also to mutants of *C. reinhardtii*. For example, to avoid any ambiguity in our chemical shift assignments, we compared the 2D spectrum of the starchless strain *sta 6-1* to that of the wild type strain (Figure 6). The main differences confirm our assignment of starch, however we observed additional correlations in whole wild type (*wt*) cells that are absent in whole starchless mutants (see the dashed lines and circled peaks in Figure 6). These correlations can be described as one spin system of six carbons (at 92.7, 72.4, 74.9, 76.9, 70.4 and 61.4 ppm from C_1 to C_6) and one correlation at 96.7 ppm/75.0 ppm. We suspect them to be co-products of starch synthesis in cells that in situ SS-NMR of various *C. reinhardtii* strains has allowed to detect. An assignment of those peaks is preliminary at best, however the chemical shifts of the six correlated carbons could correspond to those observed in high-energy twisted starch helices associated with starch synthesis in the cell [50,51].

In cells, SS-NMR can also be used to monitor cellular growth under different conditions. Indeed, we have compared various *C. reinhardtii* strains under nitrogen-rich and nitrogen-deprived diets. The effects are such that simple 1D NMR and spectral subtraction are enough to isolate starch signals. For example, after six days of cell culture, the difference spectrum between the *wt* strain and the starchless mutant (*sta 6-1*) leads to an in situ starch spectrum with a fairly good resolution (Figure 7). This type of approach can be used to determine the crystallinity of starch in the storage grains of the microalga in situ, with similar results to purified starch (see Table S1). Similarly, under nitrogen starvation of amylopectin-rich (strain *sta 3-3*) and amylose-rich (strain *st 2-1*) starch producing mutants, starch overproduction is so intense that a subtraction of its spectrum to that of microalgal cells grown in normal medium resulted in an in situ spectrum of starch (data not shown).

Although nitrogen depletion initially leads to detectable starch overproduction, starch signals were comparable to those of normal cells after one month. More than 80% of microalgal cells survived without any addition of carbon or nitrogen sources because starch reserves had become *C. reinhardtii*'s main nutrient.

Living microalgae were introduced in the NMR spectrometer and were monitored by ^{13}C SS-NMR under magic-angle spinning. During the course of the experiments, microalgae might die from a combination of spinning, heating and starving, however the structures of most of their constituents remain intact. In cells, ^{13}C SS-NMR is thus a versatile approach that enables the identification of starch in microalgae, the comparison of various strains or growth conditions and the study of algal metabolism by monitoring the amount of starch and its progressive degradation throughout the cell life. Starch accumulation, which is a well-known metabolic response to stress, suggests that whole cell SS-NMR experiments could be a useful tool to monitor stress in microalgae.

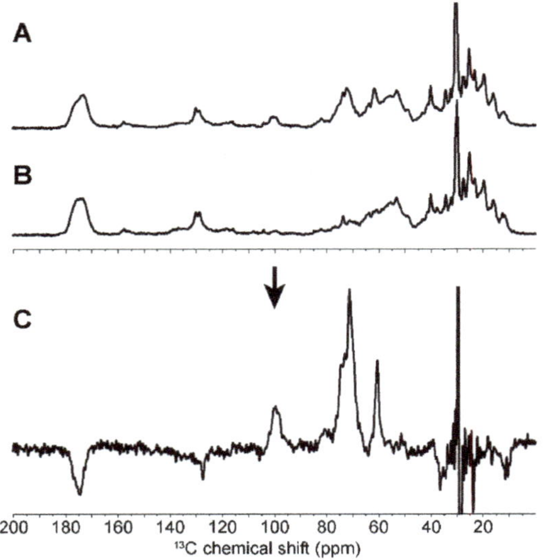

Figure 7. 1D cross-polarisation ^{13}C solid-state NMR spectra of whole *C. reinhardtii wt* cells (**A**) and *sta 6-1* starchless strain cells (**B**). Subtraction of these spectra results in the 1D spectrum of in situ native starch (**C**).

3. Materials and Methods

3.1. Materials

^{13}C-labelled (99%) sodium bicarbonate was obtained from Cambridge Isotope Laboratories (Tewksbury, MA, USA) or Martek Isotopes LLC (Olney, MD, USA). Natural abundance maize starch, all chemicals used for the growth medium, starch extraction and cell viability assays, such as Percol® and Fluorescein DiAcetate (FDA) were purchased from Sigma-Aldrich (Oakville, ON, Canada).

3.2. Strain, Media and Growth Conditions

Strains. Wild type strain of *Chlamydomonas reinhardtii 222+* was obtained from the collection of the Institut de Biologie Physico-Chimique (Paris, France). Other strains were obtained from the *Chlamydomonas Resource Center* at the University of Minnesota (http://chlamycollection.org/). Strains *sta 3-3* (CC 2916, soluble starch synthase III mutant [52]), *st 2-1* (CC 2687, granule bound starch synthase mutant [53]) and *sta 6-1* (CC 5373, ADP-glucose phosphorylase mutant [54,55]) are amylose-rich, amylopectin-rich and inhibited starch producing strains, respectively.

Algal growth. Tris minimal medium buffered with HCl to pH 7.3 was made as described by Surzycki [56]. Briefly, the medium in polyethylene Erlenmeyer (Nalgene, Thermo Scientific, Waltham, MA, USA) was inoculated with microalgae kept on TAP-medium 1.6% agar plates. The tris-minimal medium was supplemented with 1 g/L of sodium bicarbonate filtered immediately prior to cell culture. For cellular growth, heterotrophic conditions were imposed by purging CO_2 and O_2 through gas nitrogen bubbling before sealing the Erlenmeyer. Thereafter, the cells were grown under continuous white light illumination (100 µmol photons·m^{-2}·s^{-1}) at 23 °C $\pm$ 1 °C with gentle agitation (100 rpm). Five days were needed to reach the exponential phase (5.10^6 cell/mL). Cells were harvested after

$1500 \times g$ centrifugation for 10 min. For ^{13}C labelling, labelled sodium bicarbonate (NaH$^{13}CO_3$) was used as the carbon source at 1 g/L in the Tris minimal medium.

Whole cell experiments. For whole cell experiments, microalgal cells were centrifuged, then rinsed twice with a buffer containing 86 mM NaCl and ~50 mg were packed in a 3.2 mm Varian NMR rotor.

Starch overproduction. As larger volumes are needed for starch purification, a 300 mL preculture of wild type *C. reinhardtii* was grown to 4×10^6 cells/mL and then 50 mL of this solution was added to 2.5 L of medium contained in a 6 L Erlenmeyer. Again, cells were harvested in the exponential phase ($5–6 \times 10^6$ cells/mL) after 6 days. Starch overproduction in mutant strains was achieved via nitrogen starvation using the medium described by Ball et al. [22], consisting of the normal growth media simply omitting nitrogen. According to this article, nitrogen depletion is the best compromise, under heterotrophic conditions, between cell production and starch accumulation compared to phosphate and sulphur depletion. This overproduction represents a tenfold final improvement in pure starch yield.

3.3. Starch Purification

We adapted the protocols of starch extraction from Buléon and co-workers [11]. Briefly, cells were harvested in the exponential phase by centrifugation for 10 min at $1500 \times g$ and 4 °C. Cells were rinsed twice with HEPES buffer (HEPES 250 mM, MgCl$_2$ 5 mM, sucrose 300 mM, pH 7.3 with EDTA 10 mM, benzamidine 1 mM and PMSF 200 μM as antiproteases) and were then centrifuged again.

The cells pellet was diluted to 10^8 cells/mL in the same buffer and was disrupted using a homogenizer (EmulsiFlex-C5, Avestin, Ottawa, ON, Canada). Cells were passed through this cell homogenizer 4 times at 10,000 psi and were centrifuged at $2000 \times g$ for 20 min, thus pelleting big cell debris and starch granules. Starch was washed several times with Milli-q water and was further purified using Percol® (one volume starch for 4 volumes of Percol). The final pellet was considered as pure native starch and was used without further purification for XRD and NMR. For storage that was longer than 3 days, starch was lyophilized. NMR and XRD tests proved that rehydrated starch reaches the same structure than the native one. For NMR experiments, starch was hydrated and ~50 mg was packed in a 3.2 mm Varian NMR rotor.

3.4. Starch Retrogradation

A-type starch can be transformed into B-type starch using retrogradation, i.e., recrystallization of the amorphous phase after gelatinization [57]. First, starch was boiled for 15 min in excess water (10% w/w suspension) at 100 °C in a sealed glass tube. The resulting gel was then slowly cooled and stored at 4 °C. After one month, B-type XRD patterns were observed (Figure S1B), which was in good agreement with the literature [58]. This product was considered as the B-type model starch in this study.

3.5. Cell Viability

To test cell viability, fluorescence analysis is performed on a BD Accuri™ C6 flow cytometer (BD Biosciences, Mississauga, ON, Canada). A total of 1000 intact cells were acquired for each sample. This test measures the natural chlorophyll fluorescence and FDA-derived fluorescence generation based on esterase activity level [59]. Briefly, 5 μL of a 1 mM FDA solution was added to 1 mL of cell solution at 2×10^6 cells/mL for 20 min prior to measurement. Thus, cell viability was verified to be above 93% before each whole cell NMR experiment.

3.6. Solid-State NMR

All spectra were recorded on a Bruker Avance III-HD (Milton, ON, Canada) operating at a frequency of 150.87 MHz for ^{13}C and 599.95 MHz for proton (^1H) using a Varian 3.2 mm magic-angle spinning (MAS) triple resonance probe (Agilent, Santa Clara, CA, USA). The spinning frequency was set to 15 kHz and the probe was kept at room temperature corresponding to a sample temperature of approximately +35 °C for all experiments. 1D (single-pulse (SP) and cross-polarization (CP)) as well as 2D experiments (Incredible Natural Abundance Double QUAntum Technique (INADEQUATE)) were recorded with nutation frequencies of 60 kHz and 75 kHz for carbon and proton channels, respectively (corresponding to 4.2 and 3.35 µs 90° pulses) [29,30,33,60] and two-pulse phase modulation (TPPM) dipolar decoupling. For CP spectra, 256 scans were typically recorded with a recycling delay of 15 s and an optimized 1 ms contact time (total duration of 1 h). For SP, 128 scans were recorded with 30 s of recycling delay (total duration of 1 h). For 2D experiments, CP or a 2 s long Nuclear Overhauser Effect (NOE) pulse sequence were used to transfer nuclear spin polarization from ^{1}H to ^{13}C, with a 3 s recycle delay (1 s for NOE), an acquisition time of 25 ms, a total of 768 t1 increments of 32 scans and quadrature acquisition using the states-Time Proportional Phase Incrementation (TPPI) method (total duration of ~20 h). The delay τ during which the J couplings evolve was set to 2 ms. For NOE-INADEQUATE spectra, an NOE delay of 1 s was used. ^{13}C chemical shifts were externally calibrated with respect to adamantane fixing the CH_2 resonance at 38.48 ppm [61]. Spectra were processed using Topspin (Bruker) or Mnova software (Mestrelab Research, Santiago de Compostela, Spain). No line broadening was used on the 1D spectra, while the zero-filling to 1024 points and a sine squared apodization were applied in both dimensions prior to Fourier transform of the INADEQUATE spectra.

3.7. X-ray Diffraction

X-ray diffraction was performed on dry or wet samples (~130 mg). Patterns were recorded in transmission mode on a *Bruker D8 Advance* (Milton, ON, Canada) diffractometer operating at 40 mA and 40 V. The Cu K$_\alpha$ radiation (λ = 1.5406 Å) was selected using a motorized slit of 0.681 nm. Data were recorded in triplicate, with incident angles (2θ) ranging from 2° to 60°, and signals were averaged prior to normalization. Diffractograms were normalized to the same total area under the scattering curve over the Bragg angle range 5–45° (2θ).

3.8. Crystallinity Quantification

Crystallinity was assessed by SS-NMR using the method proposed by Lopez and co-workers [62] on 1D CP experiments. Briefly, a positive or negative weighting factor was applied to each resonance for purely crystalline or purely amorphous resonances, respectively. Starch crystallinities are listed in Supplementary Table S1.

4. Conclusions

In this work, we presented an unambiguous full assignment of native and retrograded starches in the A and B forms, amylose- and amylopectin-rich starches. The high resolution that was obtained in 2D ^{13}C INADEQUATE NMR experiments enabled, for the first time, the assignment of the C$_2$, C$_3$ and C$_5$ sites in different starch samples. This work confirmed differences that were observed previously between various starch types, and also added C$_3$ and C$_5$ chemical shifts and multiplicity to discriminate between them. Furthermore, full new spin systems were described for amorphous starch and non-reducing end groups of starch. ^{13}C SS-NMR spectra also allowed the assessment of starch crystallinity, disorder and dynamics, and could also be used to evaluate the chain length and amylose/amylopectin content.

This study is a step forward in the differentiation between saccharides within a microalgal cell. It is the first one reporting *in-cell* SS-NMR measurements with sufficient resolution to distinguish amorphous, A- and B-type starches, without any time-consuming and potentially altering purification steps. Moreover, we showed that starch crystallinity can be assessed in the cell. The application of this methodology to various *C. reinhardtii* mutants under different growth conditions and the detection of by-products of starch biosynthesis and its metabolism show how this approach could be applicable to the in situ study of other microorganisms. This work also represents a solid base to study physicochemical functionalisation of starch matrices and starch degradation in bioenergy, food industry or drug delivery contexts, for example.

Supplementary Materials: Supplementary materials can be found online.

Author Contributions: Methodology and analysis, A.P. and A.A.A.; writing, A.P., A.A.A., D.E.W. and I.M.; supervision, D.E.W. and I.M.; funding acquisition, I.M.

Funding: This research was funded by Natural Sciences and Engineering Research Council (NSERC) of Canada, grant number 326750-2013. The APC was funded by the same grant.

Acknowledgments: The authors would like to thank Denis Flipo (UQAM) for flow cytometry guidance, Bhushan Nagar (McGill University) for the use of their Emulsiflex, David Dewez (UQAM) for sharing his microalgal cell growth facilities and for his advice, Francesca Zito (CNRS, France) for providing the wild-type Chlamydomonas reinhardtii strain and for her advice on Chlamydomonas growth and metabolism and J.-P. Bourgouin for his help with microalgal cell culture.

Conflicts of Interest: The authors declare no conflict of interest.

Abbreviations

2D	Two dimensional
SS-NMR	Solid state Nuclear Magnetic Resonance
^{13}C	Carbon 13
μm	Micrometer
nm	Nanometer
XRD	X-ray diffraction
1D	One dimensional
g/L	Gram per liter
rpm	Rotation per minute
NaH^{13}CO$_3$	Sodium bicarbonate
mg	Milligram
mm	Millimeter
mL	Milliliter
mM	Millimole per liter
HEPES	4-(2-hydroxyethyl)-1-piperazineethanesulfonic acid
EDTA	Ethylenediaminetetraacetic acid
PMSF	Phenylmethylsulfonyl fluoride
FDA	Fluorescein DiAcetate
MHz	Mega Hertz
MAS	Magic Angle Spinning
^{1}H	Proton
kHz	Kilo Hertz
SP	Single pulse
CP	Cross polarization
INADEQUATE	Incredible Natural Abundance Double QUAntum Technique
μs	Microseconde
TPPM	Two-pulse phase modulation
NOE	Nuclear Overhauser Effect
TPPI	Time Proportional Phase Incrementation
Å	Ångström
DQ	Double quantum
ppm	Parts per million
wt	Wild type
ap	Amylopectin-rich starch
as	Amylose-rich starch
retro	Retrograded wild-type starch
am	Amorphuous starch

References

1. Pérez, S.; Bertoft, E. The molecular structures of starch components and their contribution to the architecture of starch granules: A comprehensive review. *Starch/Staerke* **2010**, *62*, 389–420. [CrossRef]
2. Sajilata, M.G.; Singhal, R.S.; Kulkarni, P.R. Resistant starch—A review. *Compr. Rev. Food Sci. Food Saf.* **2006**, *5*, 1–17. [CrossRef]
3. Paleos, C.M.; Sideratou, Z.; Tsiourvas, D. Drug delivery systems based on hydroxyethyl starch. *Bioconjug. Chem.* **2017**, *28*, 1611–1624. [CrossRef] [PubMed]
4. Bertoft, E. Understanding starch structure: Recent progress. *Agronomy* **2017**, *7*, 56. [CrossRef]
5. Sarko, A.; Zugenmaier, P. Crystal structures of amylose and its derivatives. In *Fiber Diffraction Methods*; American Chemical Society: Washington, DC, USA, 1980; Volume 141, pp. 459–482.
6. Wu Hsien-Chih, H.; Sarko, A. The double-helical molecular structure of crystalline A-amylose. *Carbohydr. Res.* **1978**, *61*, 27–40. [CrossRef]
7. Wu Hsein-Chih, H.; Sarko, A. The double-helical molecular structure of crystalline B-amylose. *Carbohydr. Res.* **1978**, *61*, 7–25. [CrossRef]
8. Popov, D.; Buléon, A.; Burghammer, M.; Chanzy, H.; Montesanti, N.; Putaux, J.L.; Potocki-Véronèse, G.; Riekel, C. Crystal structure of A-amylose: A revisit from synchrotron microdiffraction analysis of single crystals. *Macromolecules* **2009**, *42*, 1167–1174. [CrossRef]
9. Imberty, A.; Chanzy, H.; Perez, S.; Buleon, A.; Tran, V. The double-helical nature of the crystalline part of A-starch. *J. Mol. Biol.* **1988**, *201*, 365–378. [CrossRef]
10. Imberty, A.; Perez, S. A revisit to the three-dimensional structure of B-type starch. *Biopolymers* **1988**, *27*, 1205–1221. [CrossRef]
11. Buleon, A.; Gallant, D.J.; Bouchet, B.; Mouille, G.; D'Hulst, C.; Kossmann, J.; Ball, S. Starches from A to C (Chlamydomonas reinhardtii as a Model Microbial System to Investigate the Biosynthesis of the Plant Amylopectin Crystal). *Plant Physiol.* **1997**, *115*, 949–957. [CrossRef] [PubMed]
12. Rondeau-Mouro, C.; Buléon, A.; Lahaye, M. Caractérisation par RMN des biopolymères d'origine végétale, de la molécule à l'organisation supramoléculaire. *C. R. Chim.* **2008**, *11*, 370–379. [CrossRef]
13. Paris, M.B.H.; Emery, J.; Buzaré, J.Y.; Buléon, A. Crystallinity and structuring role of water in native and recrystallized starches by 13C CP-MAS NMR spectroscopy: 1: Spectral decomposition. *Carbohydr. Polym.* **1999**, *39*, 13. [CrossRef]
14. Rondeau-Mouro, C.; Veronese, G.; Buleon, A. High-resolution solid-state NMR of B-type amylose. *Biomacromolecules* **2006**, *7*, 2455–2460. [CrossRef] [PubMed]
15. Paris, M.; Bizot, H.; Emery, J.; Buzare, J.Y.; Buleon, A. NMR local range investigations in amorphous starchy substrates I. Structural heterogeneity probed by (13)C CP-MAS NMR. *Int. J. Biol. Macromol.* **2001**, *29*, 127–136. [CrossRef]
16. Paris, M.; Bizot, H.; Emery, J.; Buzare, J.Y.; Buleon, A. NMR local range investigations in amorphous starchy substrates: II-Dynamical heterogeneity probed by (1)H/(13)C magnetization transfer and 2D WISE solid state NMR. *Int. J. Biol. Macromol.* **2001**, *29*, 137–143. [CrossRef]
17. Crow, W.B. Chapter IV—The cell. In *A Synopsis of Biology*; Crow, W.B., Ed.; Butterworth-Heinemann: Oxford, UK, 1960; pp. 10–24.
18. Johnson, X.; Alric, J. Central carbon metabolism and electron transport in Chlamydomonas reinhardtii: Metabolic constraints for carbon partitioning between oil and starch. *Eukaryot. Cell* **2013**, *12*, 776–793. [CrossRef] [PubMed]
19. Rolland, N.; Atteia, A.; Decottignies, P.; Garin, J.; Hippler, M.; Kreimer, G.; Lemaire, S.D.; Mittag, M.; Wagner, V. Chlamydomonas proteomics. *Curr. Opin. Microbiol.* **2009**, *12*, 285–291. [CrossRef] [PubMed]
20. Li-Beisson, Y.; Beisson, F.; Riekhof, W. Metabolism of acyl-lipids in Chlamydomonas reinhardtii. *Plant J.* **2015**, *82*, 504–522. [CrossRef] [PubMed]
21. Blaby, I.K.; Blaby-Haas, C.; Tourasse, N.; Hom, E.F.Y.; Lopez, D.; Aksoy, M.; Grossman, A.; Umen, J.; Dutcher, S.; Porter, M.; et al. The Chlamydomonas genome project: A decade on. *Trends Plant Sci.* **2014**, *19*, 672–680. [CrossRef] [PubMed]
22. Gallaher, S.D.; Fitz-Gibbon, S.T.; Glaesener, A.G.; Pellegrini, M.; Merchant, S.S. Chlamydomonas genome resource for laboratory strains reveals a mosaic of sequence variation, identifies true strain histories, and enables strain-specific studies. *Plant Cell* **2015**, *27*, 2335–2352. [CrossRef] [PubMed]

23. Ball, S.G.; Dirick, L.; Decq, A.; Martiat, J.-C.; Matagne, R. Physiology of starch storage in the monocellular alga Chlamydomonas reinhardtii. *Plant Sci.* **1990**, *66*, 1–9. [CrossRef]

24. Gérard, C.; Colonna, P.; Buléon, A.; Planchot, V. Order in maize mutant starches revealed by mild acid hydrolysis. *Carbohydr. Polym.* **2002**, *48*, 131–141. [CrossRef]

25. Perry, P.A.; Donald, A.M. The Role of Plasticization in Starch Granule Assembly. *Biomacromolecules* **2000**, *1*, 424–432. [CrossRef] [PubMed]

26. Tang, H.R.; Godward, J.; Hills, B. The distribution of water in native starch granules—A multinuclear NMR study. *Carbohydr. Polym.* **2000**, *43*, 375–387. [CrossRef]

27. Gidley, M.J.; Bociek, S.M. Molecular organization in starches: A carbon 13 CP/MAS NMR study. *J. Am. Chem. Soc.* **1985**, *107*, 7040–7044. [CrossRef]

28. Arnold, A.A.; Bourgouin, J.P.; Genard, B.; Warschawski, D.E.; Tremblay, R.; Marcotte, I. Whole cell solid-state NMR study of Chlamydomonas reinhardtii microalgae. *J. Biomol. NMR* **2018**, *70*, 123–131. [CrossRef] [PubMed]

29. Lesage, A.; Auger, C.; Caldarelli, S.; Emsley, L. Determination of Through-bond carbon–carbon connectivities in solid-state NMR using the inadequate experiment. *J. Am. Chem. Soc.* **1997**, *119*, 7867–7868. [CrossRef]

30. Holland, G.P.; Jenkins, J.E.; Creager, M.S.; Lewis, R.V.; Yarger, J.L. Quantifying the fraction of glycine and alanine in β-sheet and helical conformations in spider dragline silk using solid-state NMR. *Chem. Commun.* **2008**, 5568–5570. [CrossRef] [PubMed]

31. Simmons, T.J.; Mortimer, J.C.; Bernardinelli, O.D.; Pöppler, A.-C.; Brown, S.P.; deAzevedo, E.R.; Dupree, R.; Dupree, P. Folding of xylan onto cellulose fibrils in plant cell walls revealed by solid-state NMR. *Nat. Commun.* **2016**, *7*, 13902. [CrossRef] [PubMed]

32. Idström, A.; Schantz, S.; Sundberg, J.; Chmelka, B.F.; Gatenholm, P.; Nordstierna, L. 13C NMR assignments of regenerated cellulose from solid-state 2D NMR spectroscopy. *Carbohydr. Polym.* **2016**, *151*, 480–487. [CrossRef] [PubMed]

33. Hong, M. Solid-state dipolar inadequate NMR spectroscopy with a large double-quantum spectral width. *J. Magn. Reson.* **1999**, *136*, 86–91. [CrossRef] [PubMed]

34. Warschawski, D.E.; Devaux, P.F. Polarization transfer in lipid membranes. *J. Magn. Reson.* **2000**, *145*, 367–372. [CrossRef] [PubMed]

35. Petersen, B.O.; Motawie, M.S.; Møller, B.L.; Hindsgaul, O.; Meier, S. NMR characterization of chemically synthesized branched α-dextrin model compounds. *Carbohydr. Res.* **2015**, *403*, 149–156. [CrossRef] [PubMed]

36. Shu, X.; Jia, L.; Gao, J.; Song, Y.; Zhao, H.; Nakamura, Y.; Wu, D. The influences of chain length of amylopectin on resistant starch in rice (*Oryza sativa* L.). *Starch/Staerke* **2007**, *59*, 504–509. [CrossRef]

37. Itoh, Y.; Crofts, N.; Abe, M.; Hosaka, Y.; Fujita, N. Characterization of the endosperm starch and the pleiotropic effects of biosynthetic enzymes on their properties in novel mutant rice lines with high resistant starch and amylose content. *Plant Sci.* **2017**, *258*, 52–60. [CrossRef] [PubMed]

38. Gregory, R.B.; Gangoda, M.; Gilpin, R.K.; Su, W. The influence of hydration on the conformation of lysozyme studied by solid-state 13C-nmr spectroscopy. *Biopolymers* **1993**, *33*, 513–519. [CrossRef] [PubMed]

39. Perry, A.; Stypa, M.P.; Tenn, B.K.; Kumashiro, K.K. Solid-state (13)C NMR reveals effects of temperature and hydration on elastin. *Biophys. J.* **2002**, *82*, 1086–1095. [CrossRef]

40. Sakellariou, D.; Brown, S.P.; Lesage, A.; Hediger, S.; Bardet, M.; Meriles, C.A.; Pines, A.; Emsley, L. High-resolution NMR correlation spectra of disordered solids. *J. Am. Chem. Soc.* **2003**, *125*, 4376–4380. [CrossRef] [PubMed]

41. Buléon, A.; Colonna, P.; Planchot, V.; Ball, S. Starch granules: Structure and biosynthesis. *Int. J. Biol. Macromol.* **1998**, *23*, 85–112. [CrossRef]

42. Srichuwong, S.; Sunarti, T.C.; Mishima, T.; Isono, N.; Hisamatsu, M. Starches from different botanical sources I: Contribution of amylopectin fine structure to thermal properties and enzyme digestibility. *Carbohydr. Polym.* **2005**, *60*, 529–538. [CrossRef]

43. Creek, J.A.; Ziegler, G.R.; Runt, J. Amylose crystallization from concentrated aqueous solution. *Biomacromolecules* **2006**, *7*, 761–770. [CrossRef] [PubMed]

44. Glaring, M.A.; Koch, C.B.; Blennow, A. Genotype-specific spatial distribution of starch molecules in the starch granule: A combined CLSM and SEM approach. *Biomacromolecules* **2006**, *7*, 2310–2320. [CrossRef] [PubMed]

45. Buleon, A.; Duprat, F.; Booy, F.P.; Chanzy, H. Single crystals of amylose with a low degree of polymerization. *Carbohydr. Polym.* **1984**, *4*, 161–173. [CrossRef]

46. Zhou, W.; Yang, J.; Hong, Y.; Liu, G.; Zheng, J.; Gu, Z.; Zhang, P. Impact of amylose content on starch physicochemical properties in transgenic sweet potato. *Carbohydr. Polym.* **2015**, *122*, 417–427. [CrossRef] [PubMed]

47. van Soest, J.J.G.; Essers, P. Influence of amylose-amylopectin ratio on properties of extruded starch plastic sheets. *J. Macromol. Sci. Part A Pure Appl. Chem.* **1997**, *34*, 1665–1689. [CrossRef]

48. Cano, A.; Jiménez, A.; Cháfer, M.; Gónzalez, C.; Chiralt, A. Effect of amylose:amylopectin ratio and rice bran addition on starch films properties. *Carbohydr. Polym.* **2014**, *111*, 543–555. [CrossRef] [PubMed]

49. Klucinec, J.D.; Thompson, D.B. Amylopectin nature and amylose-to-amylopectin ratio as influences on the behavior of gels of dispersed starch. *Cereal Chem.* **2002**, *79*, 24–35. [CrossRef]

50. Wang, S.; Yu, J.; Yu, J. Conformation and location of amorphous and semi-crystalline regions in C-type starch granules revealed by SEM, NMR and XRD. *Food Chem.* **2008**, *110*, 39–46. [CrossRef] [PubMed]

51. Bogracheva, T.Y.; Wang, Y.L.; Hedley, C.L. The effect of water content on the ordered/disordered structures in starches. *Biopolymers* **2001**, *58*, 247–259. [CrossRef]

52. Fontaine, T.; D'Hulst, C.; Maddelein, M.L.; Routier, F.; Pepin, T.M.; Decq, A.; Wieruszeski, J.M.; Delrue, B.; Van den Koornhuyse, N.; Bossu, J.P.; et al. Toward an understanding of the biogenesis of the starch granule. Evidence that Chlamydomonas soluble starch synthase II controls the synthesis of intermediate size glucans of amylopectin. *J. Biol. Chem.* **1993**, *268*, 16223–16230. [PubMed]

53. Delrue, B.; Fontaine, T.; Routier, F.; Decq, A.; Wieruszeski, J.M.; Van Den Koornhuyse, N.; Maddelein, M.L.; Fournet, B.; Ball, S. Waxy Chlamydomonas reinhardtii: Monocellular algal mutants defective in amylose biosynthesis and granule-bound starch synthase activity accumulate a structurally modified amylopectin. *J. Bacteriol.* **1992**, *174*, 3612–3620. [CrossRef] [PubMed]

54. Espada, J. Enzymic synthesis of adenosine diphosphate glucose from glucose 1-phosphate and adenosine triphosphate. *J. Biol. Chem.* **1962**, *237*, 3577–3581.

55. Zabawinski, C.; Van Den Koornhuyse, N.; D'Hulst, C.; Schlichting, R.; Giersch, C.; Delrue, B.; Lacroix, J.M.; Preiss, J.; Ball, S. Starchless mutants of Chlamydomonas reinhardtii lack the small subunit of a heterotetrameric ADP-glucose pyrophosphorylase. *J. Bacteriol.* **2001**, *183*, 1069–1077. [CrossRef] [PubMed]

56. Surzycki, S. Synchronously grown cultures of Chlamydomonas reinhardi. *Methods Enzymol.* **1971**, *23*, 67–73. [CrossRef]

57. Nocek, J.E.; Tamminga, S. Site of digestion of starch in the gastrointestinal tract of dairy cows and its effect on milk yield and composition. *J. Dairy Sci.* **1991**, *74*, 3598–3629. [CrossRef]

58. Fu, Z.Q.; Wang, L.J.; Li, D.; Zhou, Y.G.; Adhikari, B. The effect of partial gelatinization of corn starch on its retrogradation. *Carbohydr. Polym.* **2013**, *97*, 512–517. [CrossRef] [PubMed]

59. Li, J.; Ou, D.; Zheng, L.; Gan, N.; Song, L. Applicability of the fluorescein diacetate assay for metabolic activity measurement of Microcystis aeruginosa (Chroococcales, Cyanobacteria). *Phycol. Res.* **2011**, *59*, 200–207. [CrossRef]

60. Arnold, A.A.; Genard, B.; Zito, F.; Tremblay, R.; Warschawski, D.E.; Marcotte, I. Identification of lipid and saccharide constituents of whole microalgal cells by (1)(3)C solid-state NMR. *Biochim. Biophys. Acta* **2015**, *1848*, 369–377. [CrossRef] [PubMed]

61. Morcombe, C.R.; Zilm, K.W. Chemical shift referencing in MAS solid state NMR. *J. Magn. Reson.* **2003**, *162*, 479–486. [CrossRef]

62. Lopez-Rubio, A.; Flanagan, B.M.; Gilbert, E.P.; Gidley, M.J. A novel approach for calculating starch crystallinity and its correlation with double helix content: A combined XRD and NMR study. *Biopolymers* **2008**, *89*, 761–768. [CrossRef] [PubMed]

MDPI

St. Alban-Anlage 66

4052 Basel

Switzerland

Tel. +41 61 683 77 34

Fax +41 61 302 89 18

www.mdpi.com

International Journal of Molecular Sciences Editorial Office

E-mail: ijms@mdpi.com

www.mdpi.com/journal/ijms